STOCHASTIC MODELS AND OPTION VALUES

Applications to Resources, Environment and Investment Problems

CONTRIBUTIONS
TO
ECONOMIC ANALYSIS

200

Honorary Editor:
J. TINBERGEN

Editors:
D. W. JORGENSON
J. WAELBROECK

NORTH-HOLLAND
AMSTERDAM • NEW YORK • OXFORD • TOKYO

STOCHASTIC MODELS AND OPTION VALUES

Applications to Resources, Environment and Investment Problems

Edited by

Diderik LUND
SAF Centre for Applied Research
Department of Economics
University of Oslo
Blindern, Oslo, Norway

and

Bernt ØKSENDAL
Department of Mathematics
University of Oslo
Blindern, Oslo, Norway

1991

NORTH-HOLLAND
AMSTERDAM • NEW YORK • OXFORD • TOKYO

ELSEVIER SCIENCE PUBLISHERS B.V.
Sara Burgerhartstraat 25
P.O. Box 211, 1000 AE Amsterdam, The Netherlands

Distributors for the United States and Canada:

ELSEVIER SCIENCE PUBLISHING COMPANY INC.
655 Avenue of the Americas
New York, N.Y. 10010, U.S.A.

Library of Congress Cataloging-in-Publication Data

Stochastic models and option values : applications to resources,
environment, and investment problems / edited by Diderik Lund and
Bernt Øksendal.
        p.   cm. -- (Contributions to economic analysis ; 200)
    Papers presented at a conference organized by the Centre for
Applied Research at the Department of Economics, University of Oslo,
and held at Loen, Norway in 1989.
    Includes indexes.
    ISBN 0-444-88630-3
    1. Investments--Mathematical models--Congresses. 2. Finance-
-Mathematical models--Congresses. 3. Options (Finance)-
-Mathematical models--Congresses. 4. Stochastic processes-
-Congresses.   I. Lund, Diderik. II. Øksendal, B. K. (Bernt
Karsten), 1945-   . III. Universitetet i Oslo. Socialøkonomisk
institutt. Senter for anvendt forskning. IV. Series.
HG4515.2.S76 1991
332'.01'5118--dc20   91-9901
                                                                    CIP

ISBN: 0 444 88630 3

PRINTED IN THE NETHERLANDS

Introduction to The Series

This series consists of a number of hitherto unpublished studies, which are introduced by the editors in the belief that they represent fresh contributions to economic science.

The term "economic analysis" as used in the title of the series has been adopted because it covers both the activities of the theoretical economist and the research worker.

Although the analytical methods used by the various contributors are not the same, they are nevertheless conditioned by the common origin of their studies, namely theoretical problems encountered in practical research. Since for this reason, business cycle research and national accounting research work on behalf of economic policy, and problems of planning are the main sources of the subjects dealt with, they necessarily determine the manner of approach adopted by the authors. Their methods tend to be "practical" in the sense of not being too far remote from application to actual economic conditions. In addition they are quantitative.

It is the hope of the editors that the publication of these studies will help to simulate the exchange of scientific information and to reinforce international cooperation in the field of economics.

The Editors

Editors' Preface

In August 1989 the Centre for Applied Research at the Department of Economics, University of Oslo, Norway, organized a conference in Loen, Norway. The topic was the theory of investment in dynamic models of uncertainty, its foundations in mathematics, and its applications. Economists, finance theorists, mathematicians, and statisticians were invited, since it was clear that all of these fields had contributed, and would continue to contribute, to the research in this area.

The aim of the conference was to stimulate the research internationally and nationally by bringing together some of the leading people in the world with those researchers that had been active in Norway. The present book includes the majority of the papers that were presented at the conference, or, in some cases, extensions of those. We feel that the aim was reached. A fruitful exchange of ideas occurred, and we believe that these papers are of interest to a broader audience.

The organizing committee of the conference was made up of Iulie Aslaksen, Olav Bjerkholt, Erik Hernæs, and Diderik Lund. We wish to thank the organizers and the Centre's secretary, Gry Nystrøm, for making this event possible. We thank the Norwegian Research Council for Science and the Humanities for supporting the conference and the publication, and the Norwegian Research Council for Applied Social Science for supporting the conference.

A special thank to the participants and others who helped in preparing this book. In addition to the authors, we had help from Knut Aase, Steinar Ekern, Henrik Martens, Robert McDonald, and a few referees who wish to remain anonymous.

Diderik Lund and Bernt Øksendal
University of Oslo

Contents

x

I. Introduction

Stochastic Models and Option Values
D. Lund and B. Øksendal (Editors)
© Elsevier Science Publishers B.V. (North-Holland), 1991

Stochastic Models and Option Values: An Introduction

Diderik Lund

Department of Economics
University of Oslo
N-0317 Oslo, Norway

This book is a result of recent developments in several fields. Mathematicians, statisticians, finance theorists, and economists were gathered in a conference, and found several interconnections in their research. The emphasis was on common methods, although the applications were also interrelated.

The common theme is dynamic models in which uncertainty unravels over time. All but one of the papers relate to modelling of one or more agents (decision makers), who behaves optimally (in some sense) at each point in time, given his or her information set at that time.[1] This means that at any point in time, decisions are made optimally, conditional on the information available at that time. At a previous date the agent takes into account that future decisions will be made in this optimal manner.

One may find it obvious that uncertainty unravels over time and that information increases. It is not difficult, however, to find examples of economic practices that neglect the consequences for optimal behavior. Choosing among fixed production schedules by the criterion of maximizing expected profits is one. If some of the schedules will in fact be flexible, and not fixed, calculations that neglect the flexibility will underestimate such schedules. Another example is timing of an investment project by choosing the initiation date which gives the highest expected profit. In this case, it is the alternative, postponing the investment, which is flexible, and this flexibility is valuable.

Thanks to Steinar Ekern for valuable comments.

[1] The scope of the book is limited: There are no game theoretic models or models of asymmetric information.

Although many economists will agree that flexibility has a value, it may be difficult to quantify it. The remedy is to apply the methods of stochastic dynamic programming, or stochastic control.[2] Optimal behavior is described by a *strategy*, which defines the decision variable(s) at each point in time as a function of the history up to, and including, that point. A brief overview of the method in continuous time is given by Øksendal in this volume. As these mathematical tools are new, they are in development, and their applications to economic problems will certainly provide new insights in the years to come. Part III of this volume presents four papers with new results in this branch of mathematics, and some important applications to economic problems.

From an economist's point of view, the application of theories of optimal behavior is problematic if it involves estimating many unknown parameters. The traditional way to meet this problem is first, to try to reduce the number of parameters, and second, to find out whether some parameters can be observed directly as prices or price ratios in markets. Finance theory is the branch of economics which has made most progress towards this goal in relation to optimal behavior in dynamic models with uncertainty. The reason for this is, of course, that financial assets, such as shares and options, are claims to uncertain future cash flows. The crucial questions for buyers and sellers of such claims are, what are their values, and, when it comes to options, when should they be exercised.

From a mathematical viewpoint, the simplest versions of such questions may seem trivial. The optimal exercise of an American option is interpreted as an optimal stopping problem, which is a special class of stochastic control problems. Under some assumptions there exists a criterion for when to exercise the option, and an accompanying value of it. The solution depends, however, on a discount rate which is often taken as exogenous.

To provide the appropriate discount rate, or other parameters that are sufficient to solve the problem, is a central aim of finance theory. This is clearly demonstrated in part II of this book. Also, part II contains other developments in the application of finance theory to the theory of real investment.

Part IV of the book contains two papers on the exploitation of natural resources. They present different methods to deal with the inherent uncertainties over the productive capacities of such resources.

---

[2] Although the terminology is not completely standardized, "stochastic dynamic programming" is most often used about discrete-time problems, while "stochastic control" is used in continuous time, some times restricted to diffusion processes.

Financial option theory applied to real investment

The papers in this part of the book extend the existing literature on what is known as *real options*. The idea is to apply the theory of financial options, or the generalization known as *contingent claims analysis*, to the analysis of real investment problems. It is, e.g., observed that a potential real investment project, as long as it is not undertaken, is similar to a financial option which has not yet been exercised. The flexibility involved has a value in itself, which may make it optimal to postpone investment. If the value of the completed investment is uncertain, but even if it has no expected upward drift over time, it is not sufficient to observe that the value is positive today in order that immediate investment becomes optimal.

A call option is a contract between writer and buyer, which gives the buyer or a subsequent owner the right, but no obligation, to buy a specified asset at a specified price on or before a specified future date. The asset that is specified is called the *underlying asset*. An example of a real option is an undeveloped oil field. It parallels a financial call option in that its owner has the right, but no obligation (except in cases of regulation), to obtain a developed field (the underlying asset) by paying the development cost.

The set of assumptions that has been the starting point for much of the recent work on real options, is that laid out in the papers by Brennan and Schwartz (1985) and McDonald and Siegel (1986). All five papers in part II are related to this tradition. A central parameter in the models is called the *rate-of-return shortfall* by McDonald and Siegel (1984), or the *convenience yield* by Brennan and Schwartz (1985). One way to introduce this concept is to consider one unit of some exhaustible natural resource, say, a barrel of oil.

A standard assumption about spot prices of natural resources is that they grow exponentially over time. Under uncertainty, this is often translated into exponential growth of the expected price at future dates, conditional on today's price, and starting from today's price. Write this as

$$E_0(S_t) = S_0 e^{\alpha t}, \tag{1}$$

where S_t is the spot price at time t, E_0 is the expectation conditional on the information available at time 0, and α is the expected growth rate of the price.

Another well-known concept is that of a risk-adjusted discount rate. It is a discount rate that supposedly translates expected values of future cash flows of a specified (risk) class into market values today. Assume that such a discount rate can be applied to find the value at time 0 of receiving a barrel of oil at time t, and denote that rate by $\hat{\alpha}$. Then the current value must be

$$E_0(S_t)e^{-\hat{\alpha}t} = S_0 e^{(\alpha-\hat{\alpha})t} = S_0 e^{-\delta t}, \qquad (2)$$

where the last equality defines the rate-of-return shortfall, $\delta \equiv \hat{\alpha} - \alpha$.

Equation (2) is an example of a reduction in the number of parameters, from two (α and $\hat{\alpha}$) to one (δ). The same reduction is known under full certainty. If the oil price is known to grow at a constant rate, the present value of a barrel, given a constant discount rate, depends only on the *difference* between the growth rate and the discount rate.

The assumption that the two rates α and $\hat{\alpha}$ are different, is also known from the theories of resource extraction under full certainty. In the standard theory, the equilibrium price net of marginal extraction costs grows at the rate of interest, which implies that the spot price itself grows at a lower rate. This does not imply, however, that the difference has to be constant. In a market equilibrium with deposits with different extraction costs, the difference will depend on the distribution of costs and sizes of deposits, and on the demand curve.

If the price is uncertain,[3] and each deposit of the resource is regarded as similar to an infinitely lived American call option, a strictly positive value of δ is still a necessary condition for market equilibrium. If δ were not positive, it would always be optimal to postpone extraction of any deposit with positive extraction costs.

So far the discussion has considered the spot price of an exhaustible resource, with a preliminary conclusion that the rate-of-return shortfall may represent a useful reduction in the number of parameters of a valuation model. The next questions are whether this concept has its parallel in the theory of financial options, and whether the reduction of parameters also applies to the valuation of options.

The most basic option pricing theory assumes that the underlying asset is a non-dividend-paying stock, or some other financial asset which is held

---

[3] More precisely, the assumption is that the price follows a lognormal diffusion process (a geometric Brownian motion), as it does in Brennan and Schwartz (1985) and McDonald and Siegel (1986).

by investors motivated by its capital gain alone. The absence-of-arbitrage pricing theory of Black and Scholes (1973) can be explained (as in Merton (1977)) by the existence of a replicating portfolio of a riskless asset and the underlying asset, giving the same value as the option at its expiration date.[4] In order to avoid arbitrage, the value of this portfolio must be the same as the value of the option.

McDonald and Siegel (1984) pointed out that if the underlying asset does not earn a capital gain which is sufficient to motivate investors to hold it, then the replicating portfolio would also earn too little, what they called a below-equilibrium rate of return. Since no-one would want to hold the portfolio, the arbitrage argument no longer holds, and the theory has to be modified.

The problem of a below-equilibrium rate of return[5] had already been recognized in relation to options on dividend-paying stocks. Since the owner of a standard call option has no right to the stock dividends, a dividend-paying stock also earns a below-equilibrium rate of return in relation to option theory, i.e., if its dividends are disregarded and "rate of return" is interpreted as capital gains only. It is the sum of capital gains and dividends that is sufficient to induce investors to hold the stock. The assumption of a constant rate-of-return shortfall is thus completely analogous to an assumption of a continuous proportional dividend yield, and the appropriate valuation formula is found in Merton (1973).[6]

For a non-dividend-paying stock the expected price gain, α, is equal to the appropriate risk-adjusted discount rate, $\hat{\alpha}$. If it weren't, investors would not be willing to hold the stock. This explains why the parameter δ appears in the valuation models for commodities, such as extracted units of exhaustible resources, but not in the basic option pricing models.

When an asset earns a below-equilibrium rate of return, investors will not want to hold it.[7] This seems to contradict the fact that most commodities

---

[4] The replicating portfolio is equipped with an accompanying strategy, that prescribes how it should be continuously adjusted to obtain the correct terminal value. The adjustments are functions of the development of the spot price of the underlying asset, and they are self-financing.

[5] More precisely it is the *expected* rate of return which is below-equilibrium.

[6] Merton's formula applies to a European option, i.e., an option which can only be exercised at one given date. The equivalent formula for an American option, which can be exercised at any time before a given expiration date, is found as a special case of the model in McDonald and Siegel (1986) for the case when the time to expiration is infinite.

[7] The assumption is that everyone has a common information set. If someone has a different subjective probability distribution of the asset return, he or she may want to hold the asset.

are stored in some quantity. In order to explain storage of commodities
whose price returns are below-equilibrium, it is assumed that the storers
have an advantage from the storage itself. This is known as the *gross
convenience yield* of (storing) the commodity. It is reasonable to assume
that those who use the commodity as an input in production, have an
advantage of storing some of it, and it is likewise observed that governments
keep some strategic reserves.

The marginal gross convenience yield per unit stored is likely to be decreasing with the amount stored, and the storage decision will equalize this
marginal yield with the sum of (possibly increasing) marginal storage costs
and the rate-of-return shortfall, which is an opportunity cost.[8] The *net
convenience yield* (or simply, *convenience yield*) is defined as the difference
between the marginal gross convenience yield and the marginal cost of storage. From the discussion above, it is seen that *the (net) convenience yield
is equal to the rate-of-return shortfall.*

Equation (2) showed that knowledge of the rate-of-return shortfall,
under the assumption that it is constant over time, allows a translation
from today's spot price of a commodity to the value of a claim to receiving
one unit of the commodity at a future date. This has three implications,
and raises one question.

The first and obvious implication is that if the rate-of-return shortfall is
known, there is no need to estimate α and $\hat{\alpha}$ separately.

The second implication is that the value of a forward contract can be
determined as soon as the rate-of-return shortfall is known.[9] Conversely,
knowledge of the value of a forward contract determines the rate-of-return
shortfall. Thus the forward or futures markets[10] are valuable sources of
information about the sizes of rate-of-return shortfalls. One problem in the

---

[8] Cost efficiency in storage implies that the marginal storage cost is equalized among
storers. Since the rate-of-return shortfall is also the same for everyone, the marginal gross
convenience yield is in that case the same for all storers.

[9] From equation (2), the present value at time zero of receiving one unit of the commodity
at time t is $S_0 e^{-\delta t}$. The forward contract specifies an equivalent fixed sum, F_t, to be paid
for this at delivery. The present value of F_t, using the risk-free rate, is thus equal to (2),
and we find $F_t = S_0 e^{(r-\delta)t}$. The risk-free rate r is assumed to be observable.

[10] Futures and forward contracts are equivalent if the risk-free interest rate is non-stochastic.

application to real investment decisions is that traded forward or futures contracts most often are of too short maturity.

The third implication is the determination of the market value of claims to any cash flow that is a *linear function* of prices with known rate-of-return shortfalls. An example could be the (pre-tax) value of an operating profit, product prices times product quantities minus factor prices times factor quantities.[11] But then the question arises, what about cash flows that are non-linear functions of uncertain prices. One example is options.

The answer is to apply the valuation method suggested by Cox and Ross (1976).[12] This method does not look for a risk-adjusted discount rate to be applied to each cash flow. For cash flows that are complicated non-linear functions of uncertain prices with known characteristics, the calculation of an appropriate risk-adjusted discount rate (from, e.g., the Capital Asset Pricing Model) is likely to be impracticable. Instead the method prescribes that the risk-adjustment should take place in the drift term (the expected growth rate) of the price, not in the discount rate.

For linear functions the method gives the same result as the method of the risk-adjusted discount rate in equation (2). The risk adjustment is $\hat{\alpha} - r$, where r denotes the risk-free interest rate. The method says that the drift term of the price should be reduced by this amount, which gives a new (risk-adjusted) stochastic process for the price, with the same volatility as before. Then the expected value of the cash flow should be calculated conditional on this new process. Finally the present value of this should be found using a risk-free interest rate. For the claims to oil-related cash flows, the risk-adjusted drift term of the price is $\alpha - (\hat{\alpha} - r) = -\delta + r$. This shows that the reduction in the number of unknown parameters obtains even when the functions to be valued are non-linear. (It is assumed that the risk-free rate is observable.) The expected value of the risk-adjusted price is $S_0 e^{(-\delta + r)t}$. In the linear case not even the risk-free rate is needed, since the present value reduces to

$$S_0 e^{(-\delta + r)t} e^{-rt} = S_0 e^{-\delta t} . \tag{3}$$

The discussion so far gives the background for the papers in part II. The first two papers are attempts to modify the set of assumptions in Brennan

---

[11] That the market valuation operator defined on uncertain cash flows of a given future date is linear, can be shown under various sets of assumptions, see, e.g., Myers (1968) and Ross (1978).

[12] See also Constantinides (1978), Harrison and Kreps (1979), and Cox, Ingersoll, and Ross (1985).

and Schwartz (1985). They both focus on the convenience-yield interpretation of the rate-of-return shortfall. If one wants to avoid the difficulties of estimating an expected price growth rate and an appropriate risk-adjusted discount rate separately, it is attractive to try to model the convenience yield in a way which is at the same time tractable and realistic. The original assumption, that there should be a constant (relative net marginal) convenience yield, was made primarily because of tractability. These two papers show that the authors were not satisfied with the assumption, and that they have subsequently made separate attempts to improve it. Data from futures markets were used to estimate parameters and test hypotheses.

While the assumption of a constant relative yield is the easiest to deal with, the basic assumption of Brennan and Schwartz (1985) was that the absolute convenience yield could be written as a function of the spot price alone, $C(S)$ in their notation.[13] **Brennan**'s paper tests three different such models. The fourth model is called the autonomous convenience yield model. As the name suggests, it deviates and does not permit the application of known methods of contingent claims valuation.

The models are tested for seven different commodities in various time periods. The $C(S)$ models perform reasonably well for precious metals, but not for other commodities. For those, the fourth model represents an improvement in terms of realism. This implies that new theories of contingent claims analysis are needed for commodity contingent claims, except perhaps, when the commodity is a precious metal.

The fourth model also allows for a separate test of the hypothesis that the convenience yield is negatively related to the level of inventories. This hypothesis is confirmed by the data. One should be careful in interpreting this result. It is not clear whether there is causality in one direction or the other, or a simultaneous determination of inventories and convenience yield. One possible explanation goes as follows, based on the assumption of a given risk-adjusted discount rate for any given commodity price: When prices are expected to increase much in the near future, the rate-of-return shortfall is small, perhaps negative, and inventories are increased as a speculative response. When prices are expected to increase little, perhaps decrease, the rate-of-return shortfall is large, and inventories are reduced to the point where the marginal convenience yield is equally large.

The paper by **Gibson and Schwartz** is concerned with only one commodity, crude oil, and elaborates in more detail on one particular

---

[13] $C(S) = \delta S$, with δ constant, is the model with a constant relative yield, which is still the standard assumption in the literature, as in the papers by Bjerksund and by Lund.

model of the convenience yield. The model is in accordance with the $C(S)$ assumption, but adds to this a term structure of convenience yields, analogous to a term structure of interest rates. The term structure is estimated empirically on data on futures contracts, with maturities between one month and one year. It is observed that one-month forward convenience yields level off as one looks further into the future. The assumption is therefore made that the furthest forward convenience yield can be applied in the valuation of claims with longer maturities than existing futures contracts.

A second contribution of Gibson and Schwartz is a method for estimating the *"steady-state" spot price.* The usefulness of this construction is evident from the valuation expression (2) if there are short term fluctuations in spot prices.[14] The value of a deposit of a natural resource would fluctuate just as the spot price, and the criteria for investment decisions (as in Bjerksund's paper) would easily change from one day to the next.

Finally Gibson and Schwartz are able to partially test their valuation method for long term assets. There exist oil bonds issued by Standard Oil of Ohio, that can be priced according to the method. Actual market prices are also recorded, although infrequently. The model does underprice the bonds compared to the market, but a number of interesting conclusions about different versions of the model can be drawn, and suggestions for future research are made.

The contribution by **Bjerksund** is not in terms of modifying the standard assumptions in the literature, but rather in stretching the method when it comes to complexity of the claims to be valued. While the first sections of the paper apply well-known results to decision problems regarding an oil field, the later ones contain some intricate contingent claims. The

---

[14] If the assumption of the spot price following a lognormal diffusion process is maintained literally, there are no such short term fluctuations. However, Gibson and Schwartz find that they exist. One may believe that the lognormal diffusion process describes, say, monthly or yearly observations of spot prices, but that it either does not describe daily or weekly data well (because of, e.g., mean-reverting components in the very short term), or that estimates of the volatility from daily or weekly observations are higher than those estimated on monthly or yearly observations.

problem that is solved is, what does it cost a government to decide

(i)   on buying out the owner of an American call option (the oil field) at
      some future date T' at the fair market value as if the option were
      infinitely lived, conditional on the option not being exercised before
      T', and simultaneously
(ii) on exercising the option (developing the field) in case it is acquired.[15]

In the process of solving this problem, Bjerksund finds the value, Ψ, of a
claim to a cash flow at T' which is a power function of the spot price $S_{T'}$ if
this price is below a limit, but zero if it exceeds the limit.[16] He also finds the
value φ of the same power function conditional on the spot price not having
exceeded a critical boundary during an interval. In both cases these values
converge to functions that are discontinuous at $(S_{T'}, T')$. This implies that
the corresponding replicating portfolio strategies are unbounded. It is an
unresolved issue whether such unbounded strategies are an acceptable part
of an absence-of-arbitrage argument.[17] Bjerksund's final conclusions should
be in order in any case, since he only uses Ψ and φ as additive building
blocks. The cash flows that he finally wants to value, are continuous
functions of the spot price, S.

   Pindyck's contribution is more macroeconomic. He starts from the
observation that the standard neoclassical (or q) theories of investment
have poor explanatory power in macro, even in recent elaborate versions.
One aspect that is not taken care of by the theory is the full or partial
irreversibility of most investments in fixed capital. The question is whether
theories that take this into account, perform better in empirical analyses.

   Whether such theories are in the finance tradition, or result from some
other application of stochastic control or dynamic programming, a key
prediction is that higher volatility of prices leads to reduced (or postponed)
investment. The volatility in question is the one perceived for the future
at the time of investment. Although this depends on the probability of
catastrophic events that have not been observed so far, the volatility is

---

[15] Another way to put it is, what is the cost of deciding on a latest date for the exercise of
an infinitely lived American call option, but at the same time ensuring that the continuous
decision whether to exercise the option at each point in time before the latest date is made
as if the option still were infinitely lived.

[16] The reason for the interest in this, is that the value of the infinitely lived American
call option is really a power function, and that the government will have to pay this
compensation only if the spot price is below some level.

[17] See, e.g., the discussion at the end of chapter 17 of Ingersoll (1987).

estimated by sample variances, and these can be included as explanatory variables in macro investment regressions.

Pindyck performs such regressions on US data, and reports that the variance of stock returns is a highly significant explanator. One reasonable result is that its explanatory power is stronger for structures than for equipment, as the former is likely to be a more typically irreversible investment. Pindyck also shows that the magnitude of observed increases in volatility is such that it explains quite strong observed reductions in the growth rate of investment.

The paper by **Lund** compares the financial option theory with a different theory of options, originating in papers of Arrow and Fisher (1974) and Henry (1974). The Arrow-Fisher-Henry (AFH) tradition uses "option" in the general sense, and shows that flexibility has a value in itself, compared to irreversibility. While the main application of the AFH theory has been to environmental values, it is shown that the concept of flexibility value is really the same as in the financial theory.

The financial option theory can be regarded as a special case of the AFH theory, with particular assumptions on the stochastic properties of prices. Investments with environmental costs are, however, one kind of real investments that may be difficult to analyze by the financial theory, since environmental qualities are seldom traded in markets, and it may thus be difficult to establish reasonable parameter values, e.g., for the rate-of-return shortfall.

The paper suggests new applications of the financial theory to a problem that has so far only been dealt with in the AFH tradition. This has to do with the value of an uncertain consumer surplus. This is a non-marginal option, which can be valued under specific assumptions.

Stochastic control and dynamic programming

Of the four papers in this part of the book, three have a mathematical topic as their main subject, while the first one is really a paper in investment theory.

The paper by **Kobila** considers a firm whose profit depends on a stochastic parameter, Θ, e.g., demand uncertainty. The firm makes irreversible investments in capacity, K, and an optimal investment strategy is derived under the assumption that the firm maximizes expected discounted profits. The strategy is described by a concave, increasing curve in the first quadrant of the (Θ, K) plane. Starting to the left of this curve, the current

(Θ, K) pair shows only horizontal movement, as Θ is a diffusion process, while K is fixed in that part of the plane. Each time the (Θ, K) process hits the curve, investment is made so that the process stays to the left of the curve, but each time with a higher value of K.

The model provides a theory of investment which takes place in steps occuring at random times. In much of investment theory there is an assumption of convex adjustment costs to ensure that investment never becomes infinite. In this model, such ad hoc assumptions are avoided. The rate of investment per time unit is infinite, but only at scattered points in time, resulting in steps in capacity.

The model is based on an overall criterion of maximizing expected discounted profits. It is shown in the paper that a special case of the model is in accordance with a paper by Pindyck (1988). In that paper, Pindyck uses a marginal criterion for when firms should expand capacity, taking irreversibility into account. It is not completely true, however, that Pindyck's model is a special case of Kobila's, since Pindyck (1988) was able to provide a valuation of risky cash flows based on financial market equilibrium models.

In order to prove the result, Kobila proves a generalized version of the Hamilton-Jacobi-Bellman equation, which allows for singular controls. This result is of some interest in its own right. That the problems encountered in this theory requires advanced mathematical tools is also shown in an extension of the paper. Kobila (1990) compares two different models with the same kind of irreversible expansion of investment, and the same kind of demand uncertainty. One is a social planner's problem, while the other is the equilibrium of a perfectly competitive market. The price process in the latter case is described by a Skorohod stochastic differential equation[18]—not a common tool for most economists. The comparison, which is the substantive result of the paper, shows that the market is efficient.

The paper by **Brekke and Øksendal** provides a result which should have been available for a long time. The "high contact" principle has been used on several occasions to find solutions to optimal stopping problems in finance and economics. This started with Samuelson (1965) and its appendix, McKean (1965), who showed that the principle was a necessary condition for an optimum under a set of assumptions. This set of assumptions has been extended by subsequent papers, and the principle has been

---

[18] This means that it is not the usual lognormal diffusion (geometric Brownian motion), not even an Itô diffusion.

applied frequently, but with few considerations of whether the proposed solutions have really been optimal. Brekke and Øksendal provide sufficient conditions for the optimal stopping problem, ensuring that the solution is an optimum. The high contact principle is part of the sufficient conditions. To the extent that part of the literature contains results that were never proved to be optimal solutions, these conditions represent a way to complete such proofs.[19]

They also provide an application of the conditions to a simplified version of the basic model of Brennan and Schwartz (1985). In that model a firm owns a deposit of an exhaustible natural resource, say, a mine. It is costly to turn extraction on and off. The price of an extracted unit follows a lognormal diffusion. The sufficient conditions show that

(i)   the proposed strategy for opening a closed mine is optimal, given that the proposed strategy for closing an open mine is optimal, and

(ii) vice versa.

The argument also gives the value of the mine, both open and closed. However, the overall solution of the sequential problem of opening and closing the mine, is not proved to be optimal.

The **Olsen and Stensland** paper discusses a special property of some stochastic control problems. They use the term "invariant control" to describe an optimal control (optimal strategy) that remains optimal even if the instantaneous stochastic payoff is multiplied by another stochastic process. They use the same term for optimal controls that likewise retain the same functional form, but with a modification of a parameter.

The typical case is that production uncertainty affects the payoff (the reward). One solves for an optimal control, and then asks, what happens if there is price uncertainty, too. Olsen and Stensland show that under two different sets of assumptions, the control is invariant. In both sets the price process has a geometric drift component. The first set includes that the price and the other payoff factor ("production") are stochastically independent. The second does not have this independence assumption, but assumes that both the price and the production are uncontrolled lognormal diffusions.

The authors relate their result to option theory, to scheduling problems, generalizing the Gittins theorem, and to other economic problems. For the

---

[19] A similar case for establishing sufficient conditions for deterministic control problems is made by Seierstad and Sydsæter (1977) and (1983).

class of control problems considered, they show that the assumption of a geometric drift component is necessary for invariance.

The paper by **Flåm** uses a different mathematical machinery, working in discrete time. His main objective is to demonstrate the use of Lagrange multipliers in stochastic dynamic programming problems. While such multipliers are one of the best-known mathematical tools for economists, they are seldom used to solve this type of problems.

There are two types of constraints in such problems, the resource constraints and the information constraints. Flåm discusses the nature of these constraints, and the associated multipliers. He states conditions under which the multipliers exist, and presents arguments for using them in the solution of this type of problems, instead of relying on numerical methods based on a finite number of outcomes.

Statistical models of natural resource exploitation

The paper by **Berck and Johns** is of a somewhat different nature from all the others, in that it neither explicitly nor implicitly relates to a model of optimal behavior. Instead it deals with the question of how to extract information optimally.[20]

They formulate a dynamic model of a fishery, in which the stock of fish is an unobserved state variable. Their problem is to estimate the parameters of the model, and the stock of fish at different points in time. The method they use is a combination of a Kalman filter and maximum likelihood. It represents an improvement in relation to the existing literature, partly because it takes appropriate account of the measurement error.

In the application to the Pacific Halibut fishery, the estimates of stock are compared to those published by the regulatory agency. There are similarities in the time pattern, but differences in magnitude. The estimates by Berck and Johns provide better predictions of catch in the postsample period. The paper also contains suggestions for applications to other problems involving unobservable stocks.

The paper by **Stensland and Tjøstheim** has a normative purpose. It establishes an operational method for optimal decision making in the extraction from an oil field. The extraction is uncertain, modelled as a number of possible scenarios with different probabilities. As extraction

---

[20] It is reasonable to believe that the interest in their problem is motivated from the need for optimal behavior. In their particular application this can be on part of the fishing industry or on part of some regulatory agency.

proceeds, information about the realized scenario increases, and the variance of possible outcomes is reduced according to a given schedule. One specific advantage of the method is the ability to utilize any information that might be available on the properties of this schedule.

The authors show how this model can be used to find the value and the optimal extraction strategy for a given field. By numerical examples they demonstrate the value of a flexible production technology. They indicate that models with discrete time and a discrete state space can be used for finding numerical results in a variety of problems. One interesting application is to replicate the results of an analytically tractable model, e.g, in continuous time, and then alter the assumptions so that the model becomes analytically intractable. Numerical results may still be obtained, and this should serve as a useful test of the generality and robustness of analytical results.

References

Arrow K.J., and A.C. Fisher (1974): "Environmental Preservation, Uncertainty, and Irreversibility", *Quarterly Journal of Economics* 88: 312–319.

Black F., and M. Scholes (1973): "The Pricing of Options and Corporate Liabilities", *Journal of Political Economy* 81: 637–659.

Brennan M.J., and E.S. Schwartz (1985): "Evaluating Natural Resource Investments", *Journal of Business* 58: 135–157.

Constantinides G.M. (1978): "Market Risk Adjustment in Project Valuation", *Journal of Finance* 33(2): 603–616.

Cox J.C., J.E. Ingersoll, jr., and S.A. Ross (1985): "An Intertemporal General Equilibrium Model of Asset Prices", *Econometrica* 53(2): 363–384.

Cox J.C., and S.A. Ross (1976): "The Valuation of Options for Alternative Stochastic Processes", *Journal of Financial Economics* 3: 145–166.

Harrison M.J., and D.M. Kreps (1979): "Martingales and Arbitrage in Multiperiod Securities Markets", *Journal of Economic Theory* 20: 381–408.

Henry C. (1974): "Investment Decisions Under Uncertainty: The 'Irreversibility Effect'", *American Economic Review* 64(6): 1006–1012.

Ingersoll J.E., jr. (1977): *Theory of Financial Decision Making*, Totowa, NJ: Rowman and Littlefield.

Kobila T.Ø. (1990): "Equilibrium Investment and Social Efficiency under Uncertainty", Preprint, Department of Mathematics, University of Oslo.

McDonald R.L., and D.R. Siegel (1984): "Option Pricing When the Underlying Asset Earns a Below-Equilibrium Rate of Return: A Note", *Journal of Finance* 39: 261–264.

McDonald R.L., and D.R. Siegel (1986): "The Value of Waiting to Invest", *Quarterly Journal of Economics* 101: 707–727.

McKean H.P., jr. (1965): "Appendix: A Free Boundary Problem for the Heat Exchange Equation Arising from a Problem in Mathematical Economics", *Industrial Management Review* 6: 32–39.

Merton R.C. (1973): "Theory of Rational Option Pricing", *Bell Journal of Economics and Management Science* 4(1): 141–183.

Merton R.C. (1977): "On the Pricing of Contingent Claims and the Modigliani-Miller Theorem", *Journal of Financial Economics* 5: 241–249.

Myers S.C. (1968): "Procedures for Capital Budgeting under Uncertainty", *Industrial Management Review* 10: 1–19.

Pindyck R.S. (1988): "Irreversible Investment, Capacity Choice, and the Value of the Firm", *American Economic Review* 78(2): 969–985.

Ross S. (1978): "A Simple Approach to the Valuation of Risky Streams", *Journal of Business* 51(3): 453–475.

Samuelson P.A. (1965): "Rational Theory of Warrant Pricing", *Industrial Management Review* 6: 13–31.

Seierstad A., and K. Sydsæter (1977): "Sufficient Conditions in Optimal Control Theory", *International Economic Review* 18: 367–391.

Seierstad A., and K. Sydsæter (1983): "Sufficient Conditions Applied to an Optimal Control Problem of Resource Management", *Journal of Economic Theory* 31: 375–382.

Stochastic Models and Option Values
D. Lund and B. Øksendal (Editors)
© Elsevier Science Publishers B.V. (North-Holland), 1991

Stochastic Control Theory —
A Brief Summary

Bernt Øksendal

Department of Mathematics
University of Oslo
N-0316 Oslo, Norway

1. Introduction. The stochastic control problem

The purpose of this article is to give a brief introduction to stochastic control theory for the uninitiated. Basically the presentation will be in two parts:

(A) The dynamic programming principle and the Hamilton-Jacobi-Bellman (HJB) equation (sections 1–2).
(B) The assumptions for the validity of the HJB equation: Nonsmooth solutions and singular controls (sections 3–4).

Let us begin by outlining the kind of systems that stochastic control theory applies to:

The *system* that we consider can be described by an n-dimensional stochastic differential equation of the form

$$dX_t = b(t, X_t, u_t)dt + \sigma(t, X_t, u_t)\,dB_t, \qquad (1)$$

where $X_t \in \mathbf{R}^n$ and $b(t, x, v) : \mathbf{R} \times \mathbf{R}^n \times \mathbf{R}^k \to \mathbf{R}^n, \sigma(t, x, v) : \mathbf{R} \times \mathbf{R}^n \times \mathbf{R}^k \to \mathbf{R}^{n \times n}$ are given (t, x)-Lipschitz continuous functions for each $v \in \mathbf{R}^k$ and $B_t = \{B_t(\omega), t \geq 0, \omega \in \Omega\}$ denotes n-dimensional Brownian motion. We interpret $X_t = X_t(\omega)$ as the *state* of the system at time t, $u_t = u_t(\omega)$ is the *control* that we apply at time t in order to maximize the *expected performance*, $J^u(t, x)$, of the system,

$$J^u(t, x) = E^y \left[\int_0^\tau F^u(Y_s)ds + K(Y_\tau) \right]$$

$$= E^{t,x} \left[\int_t^{\tau + t\,?} F^u(s, X_s)ds + K(\tau, X_\tau) \right], \qquad (2)$$

Thanks to Olav Bjerkholt and Kjell Arne Brekke for helpful comments.

where $E^{t,x}$ denotes the expectation with respect to the probability law
$P^{t,x} = P^y$ of the process $(t+s, X_{t+s}) = Y_s$ starting at $(t,x) = y$, $F^u(t,x) = F(t,x,u)$ and $K(t,x)$ are given continuous functions with $F \geq 0$ and K
bounded and $\tau = \tau_G$ is the first exit time for Y_s of a given open G in $\mathbf{R} \times \mathbf{R}^n$:

$$\tau = \inf\{s > 0; Y_s \notin G\}. \tag{3}$$

(For example, τ could be a deterministic time $T \leq \infty$ which corresponds
to $G = (-\infty, T) \times \mathbf{R}^n$. We interpret $K(Y_\tau(\omega))$ as 0 if $\tau(\omega) = \infty$.) For a
more detailed presentation see Øksendal (1989), Ch. XI.

Different types of stochastic controls $u_t(\omega)$ may be considered. It is a
minimum requirement that the process $\{u_t(\omega)\}_{t\geq 0}$ is $\mathcal{F}_t$-*adapted*, where $\mathcal{F}_t$
is the σ-algebra or "history" generated by $B_s(\cdot); s \leq t$. This means that for
each t the random variable $u_t(\cdot)$ is $\mathcal{F}_t$-measurable. The intuitive meaning
of this is simply that the value we decide for u_t at time t depends only
upon the "history" of the Brownian motion up to that time and not upon
future events. It turns out that we can obtain the same performance by
restricting ourselves to controls of the form (with an abuse of notation)
(see Øksendal (1989), Th. 11.3.)

$$u_t(\omega) = u(t, X_t(\omega)) \quad \text{for some} \quad u : \mathbf{R} \times \mathbf{R}^n \to \mathbf{R}^k. \tag{4}$$

Such controls are called *Markov controls*, because the ensuing corresponding solution $Y_t = (t, X_t)$ of (1) is a Markov process. (In this article we
assume at the outset that b, σ and u are such that a solution of (1) exists.)

The stochastic control problem can then be formulated as follows: Let
W be a closed subset of $\mathbf{R}^k$. Find a Markov control u^* (with values in W)
which gives *maximal performance*, i.e.

$$H(y) := \sup_u J^u(y) = J^{u^*}(y); \quad y = (t,x), \tag{5}$$

the sup being taken over all Markov controls u with values in W. We call
u^* (if it exists) an *optimal* Markov control.

2. The dynamic programming principle and the Hamilton-Jacobi-Bellman (HJB) equation

One of the fundamental results in stochastic control theory is the following:

Theorem 1. (The dynamic programming principle)
For all (deterministic) $T > 0$ and all Markov controls u we have

$$J^u(y) = E^y \left[\int_0^{T \wedge \tau} F^u(Y_s)ds + J^u(Y_{T \wedge \tau}) \right], \tag{6}$$

and hence

$$H(y) = \sup_u \left\{ E^y \left[\int_0^{T \wedge \tau} F^u(Y_s)ds + H(Y_{T \wedge \tau}) \right] \right\}, \tag{7}$$

where $\tau = \tau_G$ as before. For a proof see Øksendal (1989), p. 151–152.

Intuitively (7) says that a control which is optimal locally in time $(s < T)$ based on the assumption that an optimal control is used afterwards $(s \geq T)$, actually gives a global optimal control.

Now assume that an optimal control u^* exists. Substituting $u = u^*$ in (6) gives

$$H(y) = E^y \left[\int_0^{T \wedge \tau} F^{u^*}(Y_s)ds \right] + E^y[H(Y_{T \wedge \tau})],$$

and therefore

$$\frac{E^y \left[\int_0^{T \wedge \tau} F^{u^*}(Y_s)\,ds \right]}{E^y[T \wedge \tau]} + \frac{E^y[H(Y_{T \wedge \tau})] - H(y)}{E^y[T \wedge \tau]} = 0.$$

If we let $T \downarrow 0$ the first term approaches $F(y)$ while the second term converges, if $H \in C^2$, towards

$$A^{u(y)}H(y) = \frac{\partial H}{\partial t} + \sum_i b_i(y, u(y)) \frac{\partial H}{\partial x_i}$$
$$+ \sum_{i,j} a_{ij}(y, u(y)) \cdot \frac{\partial^2 H}{\partial x_i \partial x_j}, \tag{8}$$

where $a_{ij} = (1/2)(\sigma \sigma^T)_{ij}, \sigma^T$ being the transposed of σ. $A^{u(y)}$ is called the generator of the process Y_t^u obtained by applying the Markov control u. (C^k denotes the class of k times continuously differentiable functions, $k = 0, 1, 2, \ldots$).

In this way we obtain the following:

Theorem 2. (The Hamilton-Jacobi-Bellman (HJB) equation)
Suppose the following holds:

(i)　$H \in C^2$.
(ii) An optimal control u^* exists.

Then

$$\sup_{v \in W} \{F^v(y) + A^v H(y)\} = F^{u^*(y)}(y) + A^{u^*(y)} H(y)$$

$$= 0 \quad \text{for all } y \in G, \tag{9}$$

and

$$H(y) = K(y) \quad \text{for all } y \in \partial_R G, \tag{10}$$

where $\partial_R G$ denotes the Y^{u^*}-regular points of the boundary ∂G of G. (See e.g. Øksendal (1989), Ch. IX.)

　　Theorem 2 says that equations (9) and (10) are necessary for the optimal performance function H. It is just as important to know that these equations are sufficient:

Theorem 3. (Converse of the HJB equation)
Suppose $h \in C^2(G) \cap C(\overline{G})$ satisfies

$$F^v(y) + A^v h(y) \leq 0 \quad \text{for all } y \in G, v \in W, \tag{11}$$

and

$$h(y) = K(y) \quad \text{for all } y \in \partial G. \tag{12}$$

Then

$$h(y) \geq J^u(y) \quad \text{for all controls } u.$$

Moreover, if for each $y \in G$ we can find $u_0(y)$ such that

$$F^{u_0(y)}(y) + A^{u_0(y)} h(y) = 0,$$

then

$$h(y) = J^{u_0}(y),$$

and hence $h = H$ and u_0 is an optimal control.

3. When are the assumptions of the HJB equation satisfied?

If we assume that conditions (i), (ii) of the HJB equation are satisfied, then we can proceed as follows to find H and u^*: We first fix $y \in G$ and solve (9) as a maximum problem in $v \in W$. Let $v^* = V^*(y, H, \partial H/\partial t, \partial H/\partial x_i, \ldots)$ be the value of v which gives maximum. Substituting v^* for v gives the (in general non-linear) equation in H

$$F^{v^*}(y) + A^{v^*} H(y) = 0, \quad y \in G,$$

with boundary values

$$H(y) = K(y), \quad y \in \partial_R G.$$

Although this non-linear boundary value problem may be quite a challenge in itself, it might still be justified to say that this reduces the original problem to a lower level.

However, to justify this method we must verify that assumptions (i), (ii) of Theorem 2 hold. A sufficient condition for (i) is (see Lions (1983), Th. 9):

Theorem 4. (When is $H \in C^2$?)
Suppose that the generators A^v are uniformly elliptic, uniform both in y and v, i.e.

$$\exists \lambda > 0 \text{ s.t. } \sum_{i,j} a_{ij}(y, v) \cdot \xi_i \xi_j \geq \lambda |\xi|^2 \quad \text{for all } \xi \in \mathbf{R}^n, \tag{13}$$

for all $v \in W$ and all $y \in \overline{G}$. Moreover, assume the following:

$$\sup_{v \in W} \{ \|a(\cdot, v)\|_{H^{2,\infty}} + \|b(\cdot, v)\|_{H^{2,\infty}} + \|F(\cdot, v)\|_{H^{2,\infty}} \} < \infty, \tag{14}$$

$$a(y, \cdot), b(y, \cdot) \text{ and } F(y, \cdot) \text{ belong to } C(W) \\ \text{for all } y \in \overline{G}, \tag{15}$$

$$K(\cdot) \text{ is uniformly continuous on } \partial G. \tag{16}$$

Then there exists $\alpha > 0$ such that $H \in H^{2,\infty}(G) \cap C^{2,\alpha}(G)$ and H satisfies the HJB equation.

Here

$$\|f\|_{H^{2,\infty}} = \sum_{|\alpha| \leq 2} \|D_\alpha f\|_{L^\infty}$$

denotes the Sobolev norm of order $2, \infty$, while $C^{2,\alpha}$ denotes the space of C^2-functions whose double derivatives are Hölder continuous of exponent $\alpha > 0$.

Second, for the assumption (ii) that an optimal control u^* exists, the following result is a special case of Corollary 4.8 in Haussmann and Lepeltier (1988):

Theorem 5. (The existence of an optimal control)
For each $y \in G$ define

$$M(y) = \{(a(y,v), b(y,v), z) \in \mathbf{R}^{n \cdot n + n + 1}; v \in W, z \geq 0, z \geq F(y,v)\}\,.$$

Assume the following:

$$M(y) \text{ is convex for each } y \in G\,, \qquad\qquad (17)$$

$$W \text{ is compact}\,. \qquad\qquad (18)$$

Then an optimal control u^* exists.

4. An example: Nonsmooth optimal performance and singular controls

It should be pointed out that although Theorems 4 and 5 ensure that the sufficient conditions are met in many problems there are also cases of naturally arising stochastic control problems for which the conclusions do not hold. In such cases one cannot expect the HJB equations to be valid. The control problem may still have a solution. We shall illustrate this by discussing a simple example at some length and show that it has a *singular* control of a nature that might arise in practical cases.

Example. Suppose the system Y_t is described by

$$dY_t = dY_t^u = \begin{bmatrix} dt \\ dX_t \end{bmatrix} = \begin{bmatrix} 1 \\ 0 \end{bmatrix} dt + \begin{bmatrix} 0 \\ u_t \end{bmatrix} dB_t\,, \qquad\qquad (20)$$

where B_t is 1-dimensional Brownian motion. The problem is to find the control u_t which maximizes the expected total discounted reward

$$J^u(y) = J^u(t,x) = E^{t,x} \left[\int_t^\infty f(X_s) e^{-\rho s} ds \right]\,,$$

where f is a given bounded function and $\rho > 0$ is a constant. As before we define

$$H(t,x) = \sup_u J^u(t,x).$$

The HJB equation for this problem is

$$\sup_v \left\{ f(x)e^{-\rho t} + \frac{\partial H}{\partial t} + \frac{1}{2}v^2 \cdot \frac{\partial^2 H}{\partial x^2} \right\} = 0 \quad \text{for all } x,t, \tag{21}$$

because

$$A^u = \frac{\partial}{\partial t} + \frac{1}{2}u^2 \frac{\partial^2}{\partial x^2}$$

is the generator of Y_t^u.

(21) can only hold if

$$\frac{\partial^2 H}{\partial x^2} \le 0, \tag{22}$$

and if $\partial^2 H/\partial x^2 < 0$, the corresponding optimal v is zero. Hence the HJB equation reduces to

$$f(x)e^{-\rho t} + \frac{\partial H}{\partial t} = 0, \tag{23}$$

which gives

$$H(t,x) = \frac{1}{\rho}f(x)e^{-\rho t} + \gamma(x),$$

for some function γ. From (22) we get

$$f''(x) \le 0.$$

But the given function f need not satisfy this (and since it is bounded it in fact cannot, unless it is constant). We conclude that if f is such that $f''(x) > 0$ for at least one x, then the HJB equation cannot hold. Therefore one of the assumptions (i), (ii) must fail. Which one?

To be more specific, suppose f is continuous on $[0, \infty)$, C^2 on $(0,1) \cup (1, \infty)$ and satisfies

$$f''(x) > 0 \text{ for } x < 1,$$
$$f''(x) \le 0 \text{ for } x \ge 1,$$
$$f(0) = 0, f(1) = 1, f \text{ is increasing}.$$

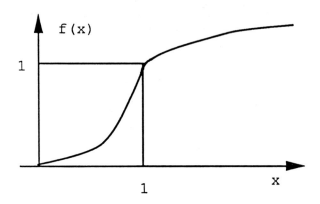

Define

$$h(t,x) = \frac{\hat{f}(x)}{\rho} e^{-\rho t}, \text{ where } \hat{f}(x) = \begin{cases} x & \text{for } x < 1, \\ f(x) & \text{for } x \geq 1. \end{cases}$$

Let us check equations (22), (23) for h:

$$\frac{\partial^2 h}{\partial x^2} \quad \begin{cases} = 0 & \text{for } x < 1, \\ \leq 0 & \text{for } x > 1, \end{cases} \tag{24}$$

$$f(x)e^{-\rho t} + \frac{\partial h}{\partial t} \quad \begin{cases} < 0 & \text{for } x < 1, \\ = 0 & \text{for } x > 1. \end{cases} \tag{25}$$

Thus HJB only holds partially for h, but we can still deduce that $h \geq H$: For any u we have

$$J^u(t,x) = E^{t,x}\left[\int_t^\infty f(X_s)e^{-\rho s}ds\right] = \int_t^\infty E^x[f(X_s)]e^{-\rho s}ds$$

$$\leq \int_t^\infty \hat{f}(x)e^{-\rho s}ds = \frac{\hat{f}(x)}{\rho}e^{-\rho t} = h(t,x),$$

because $\hat{f}$ is a superharmonic majorant of f (see e.g. Øksendal (1989), Ch. X). Hence

$$H(t,x) = \sup_u J^u(t,x) \leq h(t,x).$$

To see that in fact $H = h$ we show that we can achieve that J^u comes arbitrary close to h by choosing the control u appropriately:

For $k = 1, 2, \ldots$ define

$$u_k(t, x) = \begin{cases} k & \text{if } x < 1, \\ 0 & \text{if } x \geq 1, \end{cases}$$

and let

$$\tau_k = \inf\{s > t;\ X_s^{u_k} \notin (0,1)\}\,.$$

Then for $x < 1$ we have

$$
\begin{aligned}
J^{u_k}(t, x) &= E^{t,x}\left[\int_t^\infty f(X_s^{u_k})e^{-\rho s}ds\right] \\
&= E^{t,x}\left[\int_t^{\tau_k} f(X_s^{u_k})e^{-\rho s}ds\right] + E^{t,x}\left[\int_{\tau_k}^\infty f(X_{\tau_k}^{u_k})e^{-\rho s}ds\right]. \quad (26)
\end{aligned}
$$

From Dynkin's formula (see e.g. Øksendal (1989), Th. 7.10) we get

$$E^{t,x}[X_{\tau_k}^2] = x^2 + \frac{1}{2}k^2 E^{t,x}\left[\int_t^{\tau_k} 2ds\right] \quad \text{for } 0 < x < 1\,,$$

so

$$E^{t,x}[(\tau_k - t)] \leq \frac{1 - x^2}{k^2}\,.$$

Combined with (26) this gives

$$
\begin{aligned}
J^{u_k}(t, x) &\geq E^{t,x}\left[\int_{\tau_k}^\infty f(X_{\tau_k}^{u_k})e^{-\rho s}ds\right] - \frac{1 - x^2}{k^2} \\
&\to \frac{x}{\rho}e^{-\rho t} \quad \text{as } k \to \infty\,, \quad (27)
\end{aligned}
$$

because $E^{t,x}[f(X_{\tau_k}^{u_k})] = x$ for all k.

For $x \geq 1$ we clearly have

$$J^{u_k}(t, x) = \int_t^\infty E^{t,x}[f(X_s^{u_k})]e^{-\rho s}ds = \int_t^\infty f(x)e^{-\rho s}ds = h(t, x)\,,$$

for all k.

Thus $h = H$. In particular, H need not be a C^2 function. Moreover, in this example u^* does not exist in the usual sense. However, in view of (27) we may say that in an extended sense u^* exists and is given by

$$u^*(t,x) = \begin{cases} \infty & \text{for } x < 1, \\ 0 & \text{for } x \geq 1. \end{cases} \qquad (28)$$

with the corresponding optimal process $dY_t^* = (dt, dX_t^*)$ described as follows:

If $x < 1$ then X_t^* jumps immediately to either 1 (with probability x) or 0, and if $x \geq 1$ then $X_t^* = x$ for all t, a.s. P^x.

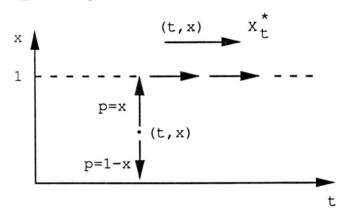

Thus we may say that the optimal behaviour Y_t^* is achieved by performing a certain singular control u^*. So in this example both conditions (i), (ii) of Theorem 2 fail.

In an article by T.Ø. Kobila (1990) it is shown that a general type of problem involving irreversible investments under uncertainty leads to singular optimal controls of the same nature as (28). For more information about singular controls, see Bensoussan and Lions (1984).

One might think that, properly formulated, it should be true in general that

$$\sup\{J^u(y); u \text{ "standard" control}\} = \sup\{J^u(y); u \text{ singular control}\}.$$

However, this is not the case. A counterexample is given by Heinricher and Mizel (1986). They generalize earlier counterexamples by Lavrentiev and Heinricher (see the references in Heinricher and Mizel (1986)).

References

Bensoussan A., and J.-L. Lions (1984): *Impulse Control and Quasi-Variational Inequalities*, Gauthier-Villars.

Haussmann V.G., and J.P. Lepeltier (1988): "On the existence of optimal controls", manuscript, to appear in *SIAM Journal of Control*.

Heinricher A.C., and V.J. Mizel (1986): "A stochastic control problem with different value functions for singular and absolutely continuous controls", Proceedings of 25th Conference on Decision and Control, Athens, Greece, December.

Kobila T.Ø. (1990): "Partial investment under uncertainty", in this volume.

Lions P.L. (1983): "On the Hamilton-Jacobi-Bellman equations", *Acta Applicandae Math.* 1: 17–41.

Øksendal B. (1989): *Stochastic Differential Equations*, Universitext, Springer-Verlag, second edition.

II. Financial Option Theory Applied to Real Investment

Stochastic Models and Option Values
D. Lund and B. Øksendal (Editors)
© Elsevier Science Publishers B.V. (North-Holland), 1991

The Price of Convenience and the Valuation of Commodity Contingent Claims

Michael J. Brennan

University of California
Los Angeles
CA 90024-1481

1. Introduction

In a recent paper Brennan and Schwartz (1985) have suggested that commodity price contingent claims, such as the ownership rights in a mine or oil well, whose payoffs depend upon the prices of storable commodities, might be valued by arbitrage methods derived from the option pricing paradigm. This approach which would obviate the need to estimate either the expected rate of change in the price of the commodity or the appropriate risk-adjusted discount rate, has obvious attractions. However, it rests upon two fundamental and so far untested assumptions. The first of these is that the stochastic process for the commodity price, in particular its variance rate, may be taken as exogenous. The second is that the convenience yield on the commodity may be written as a function solely of the spot price. In this paper we estimate and consider the predictive accuracy of three different Brennan-Schwartz type convenience yield functions, using data on futures prices for several different commodities. We also compare the Brennan-Schwartz model with an alternative model in which the convenience yield follows an autonomous stochastic process. Finally we provide some evidence for the Kaldor-Working hypothesis that the convenience yield depends on the level of inventories.

The convenience yield of a commodity is defined as the flow of services which accrues to the owner of a physical inventory but not to the owner of a contract for future delivery. Most obviously, the owner of the physical

This research which has started while I was on sabbatical leave at UCLA in 1989 has been supported by SSHRC and the Comlumbia Futures Center. I thank R. Jones for helpful discussions and D. Bensimon for research assistance. I am grateful to S. Ekern for helpful comments which have improved this version.

commodity is able to determine where it will be stored and when to liquidate the inventory. Recognizing the time lost and the costs incurred in ordering and transporting a commodity from one location to another, the marginal convenience yield includes both the reduction in costs of acquiring inventory and the value of being able to profit from temporary local shortages of the commodity through ownership of a larger inventory. The profit may arise either from local price variations or from the ability to maintain a production process despite local shortages of a raw material.

The convenience services yielded by an inventory depend upon the identity of the individual holding it; however, competition between potential storers will ensure that in equilibrium the convenience yield, net of any direct costs of storage, on the marginal unit of inventory will be equalized across all storers. Then, assuming that there exists a positive inventory of the commodity, the relation between the spot and futures price of the commodity will reflect this marginal net convenience yield in a manner which is discussed more fully below.

Kaldor (1939) appears to have been the first to recognize explicitly the role of the convenience yield of stocks or inventory in explaining the relation between concurrently observed spot and futures prices of a commodity,[1] and Working (1948, 1949) followed him in arguing that the convenience yield could explain the phenomenon of contango or negative basis observed in futures markets for agricultural commodities at certain seasons of the year. Kaldor and Working both expected the (marginal) convenience yield to depend upon the level of inventories,[2] and Brennan (1958) in estimating the risk premium element of the supply curve of storage, assumed explicitly that the convenience yield falls to zero for high inventory levels. The earliest empirical investigation of the convenience yield seems to have been conducted by Telser (1958) who found evidence of a negative relation between estimated marginal convenience yields and inventories of cotton and wheat. More recently, Fama and French (1987) have provided evidence that the marginal convenience yield varies seasonally for most agricultural and animal products but not for metal products.

Accepting that the convenience yield depends on the level of inventories, the Brennan-Schwartz assumption that it may be written as a simple

---

[1] "Stocks of goods... also have a yield *qua* stocks, by enabling the producer to lay hands on them the moment they are wanted, and thus saving the cost and trouble of ordering frequent deliveries, or of waiting for deliveries", Kaldor (1939), pp. 3–4.

[2] "The amount of stock which can thus be useful is, in given circumtances, strictly limited; their marginal yield falls sharply with an increases in the stock above requirements and may rise very sharply with a reduction in stocks below requirements", Kaldor *op.cit.* p. 4.

function of the spot price is a strong one. This would be true for example, if the only element of the net convenience yield were the cost of storage and this consisted of a fixed element and an element which depended on the unit value of the commodity such as insurance or physical deterioration; this may be a good description of the convenience yield for certain precious metals for which inventories are large and not held for the convenience services they yield at the margin. The convenience yield would also be a determinate function of the spot price if both the convenience yield and the spot price depended only on the aggregate inventory. While the former is plausible, the latter is not, unless the rate of change of the inventory depends only on the spot price and random shocks. Nevertheless, it is an empirical issue how well a model based on this simple Brennan-Schwartz assumption will perform for particular commodities.

In contrast to the Brennan-Schwartz model which posits a functional relation between the convenience yield and the spot price, the new model we develop rests on the assumption that the instantaneous rate of convenience yield follows an exogenously specified Markov process. This *autonomous convenience yield* model may be regarded as the reduced form of a more general model in which the convenience yield is determined endogenously by production, consumption, and storage decisions. A useful characteristic of the model is that in estimating it we obtain a time series of the estimated instantaneous rates of convenience yield. These convenience yield estimates are then employed to test the Kaldor-Working hypothesis that the convenience yield depends on the level of inventories. For almost all commodities we find strong evidence in favor of the Kaldor-Working hypothesis. It should be noted that our tests are more powerful than those that have been possible hitherto using the futures-spot price differential which corresponds to an integral of convenience yield rates over a period during which inventories may be expected to change.

In Section 2 the Brennan-Schwartz model and the new model with an autonomous convenience yield are presented. In Section 3 the two models are specialized to the valuation of a *convenience claim* which gives the holder the right to the net flow of convenience services over a specified time interval. Section 4 describes the data and the estimation procedure is discussed in Section 5. Empirical results are presented in Section 6.

2. Alternative valuation models

We assume, for all models, that there is a known constant rate of interest, r, and the commodity spot price, S, is determined competitively and follows the exogenously given stochastic process,

$$\frac{dS}{S} = \mu dt + \sigma(S, C) dz_s \,. \tag{1}$$

In (1) dz_s is the increment to a standard Gauss-Wiener process, C denotes the current marginal net rate of convenience yield and μ, the expected rate of price change, may depend upon an arbitrary set of state variables.

The Brennan-Schwartz model assumes that the net rate of convenience yield,[3] C, may be written as a deterministic function of the spot price, $C(S)$. This suffices to yield a determinate relation between the spot and futures prices of the commodity. Thus, let $F(S, \tau)$ denote the futures price at time t for delivery of one unit of the commodity at time T where $\tau = T - t$. The instantaneous change in the futures price is given from Ito's Lemma by

$$dF = \left(-F_\tau + \frac{1}{2}F_{ss}\sigma^2 S^2\right) dt + F_s dS \,. \tag{2}$$

Using (1) and (2), the instantaneous rate of return, including convenience yield, per dollar of investment to storage hedged in the futures market is $(SF_s)^{-1}[F_s C(S) - (1/2)F_{ss}\sigma^2 S^2 + F_\tau]$. Since the return is non-stochastic, to avoid arbitrage it must be equal to the instantaneous riskless interest rate, r, and this implies that the futures price satisfies the partial differential equation

$$\frac{1}{2}F_{ss}\sigma^2 S^2 + F_s(rS - C) - F_\tau = 0 \,, \tag{3}$$

with $F(S, 0) = S$.

Let $H(S, t)$ denote the value of a commodity price contingent claim whose payoff rate is $D(S, t)$. Using Ito's Lemma to determine the change in the value of this claim, consider the return to a portfolio consisting of a long position in the claim and a short position in (H_s/F_s) futures contracts. This return also is non-stochastic and must be equal in equilibrium to rH, the riskless return on the funds invested. Using (2) and (3) this implies that the value of the claim satisfies the partial differential equation

$$\frac{1}{2}\sigma^2 S^2 H_{ss} + H_s(rS - C) + D + H_t = rH \,, \tag{4}$$

where LHS of (4) is the return on the hedged claim position. (4) is the basic Brennan-Schwartz model for the valuation of commodity price contingent claims.

---

[3] McDonald and Siegel (1984) refer to this as the "rate of return shortfall".

The alternative model, which we shall refer to as the *autonomous convenience yield* model, assumes that the net marginal rate of convenience yield, C, follows the simple mean-reverting process

$$dC = \alpha(m - C)dt + \eta(C)dz_c, \tag{5}$$

where $\alpha > 0$ is the speed of adjustment and m is the long run mean. This assumption is motivated by the consideration that if the convenience yield is high because inventories are low, storage firms will tend[4] to have an incentive to increase their investment in inventories which, in turn, will tend to reduce the convenience yield.

Under this model the futures price will depend upon the current rate of convenience yield as well as upon the spot price and time to maturity, $F(S, C, \tau)$. Then

$$dF = \left[-F_\tau + \frac{1}{2}F_{ss}\sigma^2 S^2 + F_{sc}\rho\sigma\eta + \frac{1}{2}F_{cc}\eta^2\right]dt + F_s dS + F_c dC, \tag{6}$$

where $dz_s dz_c = \rho dt$. Consider a storage firm which invests one dollar in an inventory of the commodity, hedging its investment by shorting $(SF_s)^{-1}$ futures contracts. The return on this hedged investment, including the convenience yield is

$$
S^{-1}\left[C - F_s^{-1}\left[-F_\tau + \frac{1}{2}F_{ss}\sigma^2 S^2 + F_{sc}\rho\sigma\eta \right.\right.
$$
$$
\left.\left. + \frac{1}{2}F_{cc}\eta^2 + F_c\alpha(m - C)\right]\right]dt - (SF_s)^{-1}F_c\eta dz_c. \tag{7}
$$

The investment is not riskless because of the influence of the stochastic convenience yield on the futures price. We assume that the risk premium associated with any asset which is perfectly (positively) correlated with the stochastic change in the convenience yield is proportional to the standard deviation of the return on the asset.[5] Then the equilibrium expected return on the above portfolio may be written as $r - (SF_s^{-1})F_c\lambda\eta$, where λ is a

---

[4] Only tend to because the investment decisions of storage firms will depend upon the expected rate of change in the price of the commodity as well as upon the convenience yield.

[5] A sufficient condition for this is that the representative investor who sustains market prices have logarithmic utility and that the covariance between the convenience yield and the return on aggregate wealth be proportional to η. For a similar argument in a related context, see Brennan and Schwartz (1982).

constant of proportionality. Equating this to the drift term in (7) and rearranging we obtain the following partial differential equation for the futures price,

$$
\frac{1}{2}F_{ss}\sigma^2 S^2 + F_{sc}\rho\sigma\eta + \frac{1}{2}F_{cc}\eta^2 + F_s(rS - C)
$$
$$
+ F_c(\alpha(m - C) - \lambda\eta) - F_\tau = 0\,.
$$

(8)

Finally, let $H(S, C, t)$ denote the value of a claim to a cash flow whose rate may depend on the commodity price and the convenience yield, as well as upon calendar time, $D(S, C, t)$. The return on a portfolio formed by investing one dollar in this claim and shorting $H_s(HF_s)^{-1}$ futures contracts will be perfectly correlated with the change in the convenience yield. Arguments parallel to those advanced above then imply that the value of the claim must satisfy

$$
\frac{1}{2}H_{ss}\sigma^2 S^2 + H_{sc}\rho\sigma\eta + \frac{1}{2}H_{cc}\eta^2 + H_s(rS - C)
$$
$$
+ H_c(\alpha(m - C) - \lambda\eta) + D + H_t - rH = 0\,.
$$

(9)

Equation (9), when coupled with the appropriate boundary conditions, suffices to determine the value of any claim whose payoffs depend on the commodity price (and the convenience yield). It corresponds to the Brennan-Schwartz equation (4) when the convenience yield follows the autonomous stochastic process (5) instead of being a deterministic function of the spot price as in the Brennan-Schwartz model.

3. The convenience claim

Both the Brennan-Schwartz and autonomous convenience yield models presume knowledge of σ^2, the variance rate of the stochastic process for the spot price. We shall assume that a well developed spot market exists so that this parameter can be estimated. In addition, the Brennan-Schwartz model presumes that the parameters of the convenience yield function $C(S)$ are known, while the autonomous yield model requires knowledge not only of the parameters of the stochastic process governing the convenience yield (and its correlation with the spot price), but also of the instantaneous rate of convenience yield.

   A natural source for information on convenience yields is the commodity futures market one of whose roles is that of "price discovery". Futures markets exist for a wide range of commodities,[6] although some of the markets are of recent origin so that their history is as yet insufficient to yield accurate estimates of the parameters we require. Futures markets offer the best available price data for commodity price contingent claims, for in these markets the contracts are relatively simple and well-specified and the markets are competitive. One limitation of futures market data is the shortness of the contract maturities available relative to the maturities of the real investment projects to which the Brennan-Schwartz model was intended to apply. A second drawback is that futures contracts have payoffs which are linear in the spot price,[7] whereas many real assets yield cash flows which are nonlinear functions of the commodity spot price because output quantities may be varied in response to price. Despite these limitations we shall work with futures price data.

Since our primary focus is on the convenience yield, it is helpful to isolate that component of the futures price which represents the price of convenience. To this end we define a *convenience claim* as a claim to the net (of storage costs) flow of services yielded by a unit of inventory over a specified time period. The convenience claim is, in effect, a contract under which the holder leases a unit of inventory for a specified period. The price he is willing to pay for the lease depends on the flow of services the commodity asset will yield over the term of the lease.

Purchase of a convenience claim of maturity τ is equivalent to a spot purchase of one unit of the commodity, offset by a short position in a futures contract of maturity τ.[8] Thus the price of a τ-period convenience claim at time $t, Q(t,\tau)$ is defined by

$$Q(t,\tau) = S(t) - F(t,\tau)e^{-R(t,\tau)\tau}, \tag{10}$$

where $S(t)$ is the spot price, $F(t,\tau)$ is the futures price and $R(t,\tau)$ is the interest rate on a τ period loan at time t.

Since the convenience claim has no value at the end of the lease period we have the boundary condition

$$Q(t,0) = 0. \tag{11}$$

---

[6] Futures markets currently exist in the United States for copper, aluminum, gold, platinum, palladium, silver, crude oil, heating oil, gas oil, gasoline, lumber and propane as well as for agricultural commodities.

[7] This is not true of the recently developed commodity futures options.

[8] Throughout this paper we ignore the distinction between forward and futures contracts. See for example Cox, Ingersoll and Ross (1981).

Under the Brennan-Schwartz model $Q(t,\tau) = Q(S(t),\tau) = Q(S,\tau)$, and the value of the claim satisfies the partial differential equation (4) with $D = C(S)$. The value of the claim depends on the precise form of the convenience yield function, $C(S)$, and we shall consider three variants, as follows.

Model I: $C(S) = cS$.
Under this, the simplest model of the convenience yield, the value of the convenience claim is given by

$$Q(t,\tau) = S(t)\left(1 - e^{-c\tau}\right). \tag{12}$$

Model II: $C(S) = a + bS + cS^2$.
This function was chosen because of the additional flexibility it affords. Under this specification there does not exist a closed form solution to the valuation equation (4) which must be solved numerically.

Model III: $C(S) = \max(a, b + cS)$.
This form of the convenience yield function was motivated by the observation that so long as a commodity is held in inventory, its net convenience yield is bounded from below by the (negative of the) cost of storage: since this is likely to be constant over a wide range of inventory levels for both high value commodities and for those which can be stored in general purpose buildings, this places a lower bound on the net convenience yield, at least for these commodities. On the other hand, the spirit of the Brennan-Schwartz model suggests that the convenience yield may be higher at high spot prices (low inventory levels).

Model IV: $dC = \alpha(m - C)dt + \eta(c)dz_c$.
Under the autonomous convenience yield model the value of the convenience claim satisfies the partial differential equation (9) subject to the boundary condition (11). However, since neither the boundary condition nor the convenience yield depend upon the spot price, $Q_s = 0$ and (9) simplifies to

$$\frac{1}{2}Q_{cc}\eta^2 + Q_c[\alpha(m - C) - \lambda^*] + C - Q_\tau - rQ = 0, \tag{13}$$

where $\lambda^* = \lambda\eta$.

The solution to (13) subject to the boundary condition (11) is, for $r > 0$,

$$Q(C, \tau) = \frac{m - \lambda^*/\alpha}{r} \left(1 - e^{-r\tau}\right) - \frac{m - \lambda^*/\alpha - C}{\alpha + r} \left(1 - e^{-(\alpha + r)\tau}\right). \quad (14)$$

For $r = 0$ the solution may be written

$$Q(C, \tau) = (m - \lambda^*/\alpha)\tau - \frac{m - \lambda^*/\alpha - C}{\alpha} \left(1 - e^{-\alpha\tau}\right). \quad (15)$$

It can be seen from equation (10) that the difference between the spot price of a commodity and the convenience price for a given maturity is equal to the discounted futures price for that maturity. But the discounted futures price is simply the present value of one unit of the commodity deliverable at the maturity of the contract. Hence, given the current spot price, knowledge of the convenience price enables one to compute the present value of a unit of the commodity for future delivery; thus errors in determining convenience prices are precisely equal to the errors in the corresponding present value estimates. Therefore, by evaluating the ability of the models to predict convenience prices we are also evaluating their ability to predict the present values of futures cash flows which are linear functions of commodity prices, and therefore their potential usefulness in the capital budgeting contexts proposed by Brennan and Schwartz.[9]

4. Data

At the end of each month from January 1966 to December 1984 futures prices were collected from the *Wall Street Journal* for all of the commodities listed in Table 1, except copper for which data were taken from a tape supplied by the Columbia Center for Futures Research. The price of the immediately maturing futures contract was used as a proxy for the spot price. Consequently, convenience prices could be computed only for

---

[9] It is worth noting that while we shall be able to present standard errors of estimates only for relatively short term contracts, there is virtually no empirical evidence available for assessing the accuracy of present value estimates derived from more traditional discounted cash flow approaches.

Table 1: Futures prices sample.

Commodity	Exchange	Sample Period	Number of Futures Prices
Gold	COMEX	January 1976-December 1984	1150
Silver	COMEX	February 1976-December 1984	2106
Platinum	NYME	December 1979-December 1984	287
Copper	COMEX	January 1966-December 1984	1788
No. 2 Heating Oil	New York Mercantile Exchange	January 1972-December 1984	404
Lumber	Chicago Mercantile Exchange	January 1972-December 1984	404
Plywood	Chicago Board of Trade	January 1976-August 1983	267

those months in which there existed an immediately maturing contract. Estimation of Model IV required a regular series of convenience prices and the sample periods were selected to ensure that there existed a spot price for every month.[10]

To compute the convenience prices the end of month daily yields to maturity on the U.S. Government Bond or Bill maturing closest to the maturity of the outstanding futures contracts were taken from the CRSP U.S. Government Bond price tape. The convenience price corresponding to each outstanding futures contract was calculated from equation (10) using these yields. Finally, all spot and convenience prices were converted to 1967 dollars using the Consumer Price Index.

---

[10] We were not quite successful in this; for silver, 1 spot price and for platinum, 2 spot prices had to be *filled in* by linear interpolation from the prices of short term futures. In addition, the reported spot price for silver was used in 9 months when there was no maturing future. For plywood and lumber, futures matured only in alternate months and only the alternate month data were used in the estimations.

5. Estimation

The valuation models were estimated under the assumption that the real interest rate, r, is equal to zero. Then for the Brennan-Schwartz type models (Models I-III) the parameters to be estimated are σ^2, the variance rate of the spot price process, and a, b, c, the parameters of the relevant convenience yield function. σ^2 was estimated directly from the spot price data for each sample period under the assumption that it is an intertemporal constant.

Then consider any one of the Brennan-Schwartz type models and let δ represent the vector of parameters to be estimated. Let $g_{t\tau}(\delta)$ denote the predicted value of a convenience claim of maturity τ when the spot price is $S(t)$. For Model I $\delta = c$ and $g_{t\tau}(c) = S(t)(1 - e^{-c\tau})$. For Models II and III $\delta = \{a, b, c\}$, and $g_{t\tau}(\delta)$ is $Q(S, \tau)$ the solution for $S = S(t)$ of the partial differential equation

$$\frac{1}{2}\sigma^2 S^2 Q_{ss} + Q_s(rS - C(S)) + C(S) - Q_\tau - rQ = 0, \qquad (16)$$

subject to $Q(S, 0) = 0$ and the appropriate specification of the convenience yield function, $C(S)$. Equation (16) was solved by an implicit finite difference approximation.[11] Letting $Q_{t\tau}$ denote the observed convenience price for maturity τ at time t, the model prediction error for given δ, $\epsilon_{t\tau}(\delta)$, is given by $Q_{t\tau} - g_{t\tau}(\delta)$.

Let $\epsilon_t(\delta)$ denote the vector of prediction errors at time t and assume that $E[\epsilon_t \epsilon_{t'}] = 0$ for $t \neq t'$. Then consider an estimator of δ, $\hat{\delta}(V)$, which minimizes

$$L(\delta, \underline{V}) = \sum_t \epsilon_t(\delta)' V_t^{-1} \epsilon_t(\delta),$$

where $\underline{V} = \{V_t\}$ and V_t is a positive definite matrix. If $V_t = \Omega_t$, the variance covariance matrix of period t disturbances, $\hat{\delta}(\underline{V})$ is the maximum likelihood estimator, and $\hat{\delta}$ has a covariance matrix $(\Sigma_t X_t' \Omega_t^{-1} X_t)^{-1}$ where

$$X_t(\delta) = \left[\frac{\partial g_{t\tau}(\delta)}{\partial \delta_j} \right]$$

is a matrix. Since Ω_t is not known, an iterative estimation procedure was employed. First $\hat{\delta}(\underline{V})$ was calculated for V_t equal to the identity matrix. The residuals from this regression were then used to compute

---

[11] See Brennan and Schwartz (1978).

$\hat{\triangle}$, a diagonal matrix of the estimated residual variances. $\hat{\delta}\left(\hat{\triangle}\right)$ was then computed where $\hat{\triangle} = \left\{\hat{\triangle}_t\right\}$ and $\hat{\triangle}_t$ is the submatrix corresponding to the contract maturities outstanding in period t; the residuals from this second regression were used to calculate $\hat{\Omega}$, the estimated variance-covariance matrix of contemporaneous disturbances.[12] $\hat{\delta}(\hat{\Omega})$ was then calculated, where $\underline{\hat{\Omega}} = \{\hat{\Omega}_t\}$ and $\hat{\Omega}_t$ is the submatrix corresponding to the contract maturities outstanding in period t. Malinvaud (1964) shows that under certain regularity conditions this procedure yields an asymptotic maximum likelihood estimator.

The minimization of (16) was carried out using the Gauss-Newton method.[13] This requires that the first derivatives of the model values with respect to the model parameters δ be computed at each iteration. For Model I an analytic derivative may be computed directly from (12), the expression for the convenience price. For Models II and III it was necessary to compute the derivatives numerically; this was done by resolving the partial differential equation (14) for slight perturbations of each of the parameters a, b, c in turn, and computing the derivatives numerically.[14]

Turning next to the autonomous convenience yield model, Model IV, $g_{t\tau}(\delta)$, the predicted value of the convenience claim when the interest rate is zero, is given by Equation (15), and $\delta = (\alpha, m, \lambda^*)$. However, α and m are also parameters of the stochastic process for the convenience yield, (5). Further complication is introduced by the fact that the instantaneous rate of convenience yield is an unobservable or latent variable.[15]

For purposes of estimation the variance rate of the convenience yield is taken to be a constant, η^2. Then the exact discrete model corresponding to (5) is

$$C(t) = m\left(1 - e^{-\alpha}\right) + e^{-\alpha}C(t-1) + \xi_t, \qquad (17)$$

---

[12] The dimension of $\hat{\Omega}$ is equal to the maximum maturity of the convenience contracts. Since futures contracts of all maturities are not outstanding each month, convenience prices and prediction errors can be calculated for only a subset of the maturities in any given month. Consequently, the matrix of residuals contains many missing observations. To ensure that the estimated covariance matrix was positive definite, the missing observations were first filled in using multiple regression as suggested by Dagenais (1978).

[13] See Quandt (1983).

[14] Symmetric derivatives were calculated. See Quandt (1983).

[15] The model bears a direct resemblance to a latent variable model of the term structure of interest rates estimated by Jacobs and Jones (1985). For a general discussion of latent variable models see Aigner et al. (1983).

where $\xi_t \sim N(0, \eta^2)$. Let e_t denote the vector of valuation errors in period t augmented by the stochastic process error, ξ_t,

$$e_t = \begin{pmatrix} \xi_t \\ \epsilon_t \end{pmatrix} = \begin{bmatrix} C(t) - m(1 - e^{-\alpha}) - e^{-\alpha}C(t-1) \\ Q_t - g_t(\delta, C(t)) \end{bmatrix} \tag{18}$$

In (18) ϵ_t is the vector of valuation errors, and Q_t is the vector of observed convenience prices, $g_t(.)$ is the vector of model prices in period t, which depends on both the vector of model parameters δ and the current rate of convenience yield $C(t)$. If $E[e_t e_t'] = \Omega_t$ and $E[e_t e_{t'}'] = 0$ for $t \neq t'$, the maximum likelihood estimator $(\delta, \underline{C})$ would minimize

$$L(\delta, \underline{C}, \underline{\Omega}) = \sum_t e_t'(\delta, \underline{C})\Omega_t^{-1}e_t(\delta, \underline{C}), \tag{19}$$

where $\underline{\Omega} = \{\Omega_t\}$. Since Ω_t is not known a three step procedure similar to that described above was employed: the parameters were estimated first under the assumption that the errors were *iid*; the residuals from this regression were used to compute a diagonal variance covariance matrix; this diagonal matrix was used to compute a weighted regression and the residuals from this regression were used to estimate Ω; final parameter estimates were obtained using this estimated covariance matrix. The minimizations at each stage were again carried out using the Gauss-Newton method, the partial derivatives of (19) being computed analytically using Equations (15) and (18). This procedure yields an asymptotic maximum likelihood estimator. The instantaneous rates of convenience yield, $C(t)$, are latent variables or incidental parameters and it is well known that such unobservable variables may cause parameter estimates to be biased and inconsistent.[16] However, if there is sufficient prior information to identify the model, consistent estimation is possible. As shown in Appendix A, the model in this case is actually overidentified, even when no restrictions are placed on the contemporaneous error covariance matrix. Then the maximum likelihood procedure provides consistent and efficient estimates of the structural parameters.[17]

---

[16] See Neyman and Scott (1948).

[17] See Judge et al. (1980).

Table 2: Parameter estimates for Brennan-Schwartz convenience yield functions. Model I: $c = cS$. Model II: $c = a + bS + cS^2$. Model III: $c = \max(a, b + cS)$ (Standard errors in parentheses).

	Model I	Model II			Model III		
	c	a	b	c	a	b	c
Gold ($/ounce)							
January 1976-December 1984	-0.91×10^{-3} (0.09)	-0.17 (0.03)	4.06×10^{-3} (0.29)	-0.22×10^{-4} (0.05)			
January 1976-June 1980	-1.25 (0.16)	0.08 (0.02)	0.45 (0.25)	0.14 (0.00)			
July 1980-December 1984	-0.68 (0.08)	-0.34 (0.04)	6.10 (0.46)	-0.25 (0.01)			
January 1976-December 1984 (excluding September 1979-December 1984)	-0.14 (0.05)						
Silver (¢/ounce)							
February 1966-December 1984	-0.13×10^{-3} (0.08)	0.45 (0.06)	0.65×10^{-3} (0.17)	0.46×10^{-6} (0.00)			
February 1966-June 1975	-1.7 (1.0)	1.09 (0.08)	-13.45 (0.56)	26.67 (0.77)			
July 1975-December 1984	-0.18 (0.12)	-0.78 (0.12)	1.93 (0.27)	0.03 (0.03)			
February 1966-December 1984 (excluding September 1979-May 1981)	-0.21 (0.05)						

Table 2: (Continued) Parameter estimates for Brennan-Schwartz convenience yield functions. Model I: $c = cS$. Model II: $c = \max(a, b + cS)$. Model III: $c = a + bS + cS^2$. (Standard errors in parentheses).

	Model I	Model II			Model III		
	c	a	b	c	a	b	c
Platinum ($/ounce)							
December 1979-December 1984	-0.02×10^{-3} (0.20)	0.93 (0.12)	-8.36×10^{-3} (0.98)	14.41×10^{-6} (1.53)			
May 1981-December 1984 (0.13)	1.56 (0.13)						
Copper (¢/pound)							
January 1966-December 1984	6.05×10^{-3} (0.20)	0.48 (0.02)	-3.96×10^{-2} (0.16)	0.78×10^{-3} (0.02)	0.01 (0.00)	-2.06 (0.09)	0.05 (0.00)
January 1966-June 1975	11.45 (0.66)	0.56 (0.04)	-4.48 (0.32)	0.83 (0.05)	-1.15 (u)	-2.13 (0.09)	0.06 (0.00)
July 1975-December 1984	1.55 (0.15)	-4.03 (0.28)	13.14 (0.99)	-0.72 (0.08)	0.00 (0.00)	-1.43 (0.25)	0.04 (0.00)
No. 2 Heating oil (¢/gallon)							
September 1980-December 1984	5.49×10^{-3} (0.70)	-1.21 (0.51)	1.17×10^{-1} (0.34)	-2.24×10^{-3} (0.52)	0.23 (u)	-0.42 (0.29)	2.6×10^{-2} (1.0)
Plywood ($/thou.sq.ft.)							
January 1976-August 1983	4.26×10^{-3} (0.52)	-1.00 (0.47)	0.67×10^{-2} (1.12)	0.98×10^{-4} (0.69)	-0.05 (0.03)	-1.90 (0.15)	2.48×10^{-2} (1.70)
Lumber ($/thou.bd.ft.)							
January 1972-December 1984	1.92×10^{-3} (0.92)	3.12 (0.57)	-0.13 (0.02)	1.1×10^{-3} (0.13)	-0.84 (0.01)	-19.89 (0.58)	2.07×10^{-1} (0.05)

u = undefined

Michael J. Brennan

Table 3a: Convenience Prices for Gold (1967 $/ounce). Average
number of contracts/month 10.7.

Maturity (months)	# Obs	Convenience Prices			Standard Error of Model Predictions				
		Mean	Max	Min	Std. Dev.	I	II	IV	I[a]
Spot	108	133.15	292.24	60.70	51.98				
1	104	−0.15	1.15	−7.02	0.78	0.77	0.76	0.68	0.81
2	76	−0.31	1.15	−7.11	1.04	1.01	0.98	0.76	0.97
3	55	−0.46	1.45	−7.28	1.29	1.23	1.14	0.72	1.16
4	53	−0.31	7.05	−6.17	1.57	1.50	1.28	1.08	0.42
5	55	−0.63	2.51	−7.54	1.65	1.51	1.28	0.65	1.27
6	53	−0.77	1.40	−9.16	1.79	1.58	1.05	0.47	0.51
7	55	−0.87	2.62	−8.80	2.16	1.95	1.52	0.60	1.46
8	53	−0.89	1.50	−11.78	2.10	1.82	1.00	0.30	0.66
9	55	−1.10	2.17	−11.42	2.49	2.19	1.56	0.32	1.39
10	53	−1.09	1.60	−14.26	2.57	2.24	1.24	0.27	0.83
11	55	−1.27	2.29	−13.02	2.88	2.52	1.78	0.41	1.56
12	53	−1.32	3.96	−16.33	3.15	2.73	1.56	0.86	1.14
13	54	−1.41	2.52	−14.83	3.36	2.93	1.91	0.39	1.60
14	52	−1.31	2.41	−18.19	3.37	2.92	1.47	0.22	1.03
15	54	−1.56	2.38	−17.03	3.70	3.20	2.00	0.21	1.74
16	45	−1.66	2.88	−20.19	3.96	3.45	1.81	0.16	1.27
17	44	−2.10	2.06	−17.46	4.10	3.57	2.30	0.57	1.98
18	44	−1.77	3.59	−20.85	4.21	3.70	2.13	0.47	1.51
19	44	−2.19	2.69	−19.96	4.48	3.91	2.48	0.70	2.02
20	44	−1.78	3.95	−21.62	4.43	3.92	2.48	0.60	1.56
21	40	−1.91	2.55	−14.62	3.95	3.49	2.79	0.78	2.27
22	18	−3.02	1.37	−18.98	5.06	4.39	2.70	1.75	1.10

[a] Excluding September 1979 – May 1981

Table 3b: Convenience Prices for Silver (1967 ¢/ounce). Average
number of contracts/month 9.3.

Maturity (Months)	# Obs.	Convenience Prices			Standard Error of Model Predictions				
		Mean	Max	Min	Std. Dev.	I	II	IV	I[a]
Spot	226	280.06	1498.48	106.37	194.53				
1	181	−0.45	259.80	−188.99	24.30	24.31	24.23	16.60	3.91
2	144	−0.07	259.77	−48.08	22.55	22.58	22.29	11.09	6.59
3	116	−2.87	7.45	−198.26	18.41	18.60	18.70	7.58	1.58
4	114	−0.07	267.39	−50.36	25.73	25.79	25.13	6.25	5.81
5	115	−3.51	12.35	−206.87	19.29	19.55	19.66	5.42	2.62
6	115	−0.08	274.33	−52.09	26.32	26.41	25.39	4.31	6.08
7	115	−2.19	7.08	−23.63	4.30	4.69	4.71	3.46	2.87
8	118	−1.46	282.70	−205.43	32.44	32.54	31.56	1.82	3.13
9	116	−4.01	25.58	−207.31	20.35	20.66	20.65	2.20	6.50
10	119	−2.63	33.79	−56.46	8.62	8.98	8.20	1.70	6.77
11	116	−1.26	288.02	−208.84	33.55	33.69	32.23	1.08	3.88
12	109	−0.30	289.85	−57.63	30.00	30.21	27.99	4.98	8.98
13	103	−4.07	39.89	−207.90	21.45	21.74	21.60	1.17	4.54
14	105	−0.19	295.93	−57.95	31.35	31.62	28.81	3.06	7.96
15	103	−4.14	51.67	−208.34	22.52	22.80	22.58	3.00	5.10
16	91	0.29	302.31	−58.20	34.47	34.81	31.28	4.63	8.67
17	44	−4.00	62.97	−208.00	34.66	34.83	34.54	6.22	4.57
18	44	6.28	308.54	−60.58	50.12	51.13	45.51	8.35	11.58
19	44	0.14	77.88	−39.45	18.54	18.85	16.40	9.62	6.68
20	42	1.43	314.62	−209.47	59.63	60.01	56.12	5.49	5.76
21	41	−2.36	123.72	−203.48	40.99	41.19	39.36	14.67	12.14
22	11	−2.75	8.36	−16.30	7.36	7.29	9.78	3.80	4.60

[a] Excluding September 1979 – May 1981

6. Empirical results

(i) The Brennan-Schwartz Models

If the Kaldor-Working theory of the convenience yield is correct, and if the inventory of a commodity is sufficiently large that it is held in part for speculative purposes by individuals with no commercial interest in the commodity, then we should expect that the gross rate of convenience yield for the commodity would be zero and that the net convenience yield would be the negative of the cost of storage. Casual empiricism suggests that the precious metals, particularly gold and silver, are held for speculative purposes and that their storage costs are very low in relation to their value, so that we should expect their net convenience yields to be negative but close to zero. On the other hand, inventories of the other commodities would appear to be held mainly for commercial purposes, and therefore we should expect their net convenience yields to be mainly positive.

The simplest model of the convenience yield we have considered is Model I which assumes that the instantaneous rate of convenience yield is proportional to the spot price. Table 2 presents the maximum likelihood estimates of c, the constant proportional rate of convenience yield of Model I for different commodities and sample periods: the rates of convenience yield are monthly. The estimates for gold, silver and platinum are uniformly negative and small, the total sample estimates for these three precious metals corresponding to –1.09%, –0.16%, and –0.02% per year respectively. For the remaining "commercial" commodities, the estimated rates of convenience yield reported in Table 2 are all positive ranging from the equivalent of 2.3% per year for lumber to 7.3% per year for copper.

However Table 2 also reveals considerable intertemporal variation in the estimated rates of convenience yield for the same commodity: this is inconsistent with the assumption underlying Model I, and the variability in convenience yields is reflected in the convenience price prediction errors for Model I which are reported in Table 3 along with summary statistics on the distributions of the convenience prices. For all of the commodities the standard errors of the price predictions of Model I are very close to the standard deviations of the convenience prices themselves. For gold and platinum the standard errors of the convenience price predictions[18] are of the order of 3% of the mean spot price; for silver 12% and for the other commodities 6–10%.

---

[18] These are the same as the errors in estimating the present values of one unit of the commodity deliverable at the maturity of the convenience claim contract.

Table 3c: Convenience Prices for Platinum (1967 $/ounce). Average number of contracts/month 4.7.

Maturity (Months)	# Obs	Convenience Prices			Standard Error of Model Predictions				
		Mean	Max	Min	Std. Dev.	I	II	IV	I[a]
Spot	61	160.57	384.94	91.19	64.67				
1	34	-0.56	2.10	-13.35	2.39	2.46	2.47	2.20	2.86
2	21	-0.20	1.62	-4.34	1.18	1.20	1.09	1.11	0.54
3	21	0.10	9.14	-3.37	2.87	2.87	2.91	1.30	1.98
4	20	-0.43	1.81	-12.26	2.93	2.96	3.03	1.72	3.52
5	20	0.49	3.38	-5.33	1.96	2.02	1.85	0.73	0.92
6	21	0.32	15.60	-5.93	4.38	4.39	4.66	0.95	2.12
7	20	-0.29	2.94	-12.30	3.42	3.43	3.44	1.08	3.72
8	20	0.73	6.98	-6.65	2.94	3.04	2.75	0.20	0.89
9	21	0.53	21.72	-8.55	5.93	5.95	5.96	0.24	2.20
10	20	0.23	5.23	-11.69	4.07	4.07	4.00	0.07	3.83
11	20	2.67	12.02	-7.67	4.24	4.40	3.98	0.59	1.12
12	20	0.70	26.64	-10.89	7.34	7.37	7.31	1.09	2.12
13	15	-0.01	6.03	-10.56	5.13	5.13	4.92	1.15	4.42
14	9	1.97	9.74	-7.13	5.14	5.51	4.56	1.04	2.88
15	5	0.85	4.37	-2.02	2.21	2.37	3.15	0.22	2.97

[a] Excluding September 1979 – May 1981

Models II and III allow the convenience yield to depend upon the spot price in more general ways, and we might expect this increased flexibility to result in better price predictions. The parameter estimates and standard errors of prediction for these models are also reported in Tables 2 and 3. The Model III parameter estimates failed to converge for any of the precious metals. Model II yields significantly improved price predictions for gold but not for silver or platinum. However, the estimated convenience yield function for gold behaves perversely, for it implies that the convenience yield decreases with the spot price above $92/oz. whereas we had expected that insofar as the convenience yield depended on the spot price it would

Table 3d: Convenience Prices for Copper (1967 ¢/pound). Average
number of contracts/month 7.8.

Maturity (Months)	# Obs	Convenience Prices				Standard Error of Model Predictions			
		Mean	Max	Min	Std. Dev.	I	II	III	IV
Spot	228	41.62	96.32	17.88	15.02				
1	183	0.12	7.33	−0.29	0.65	0.63	0.59	0.59	0.56
2	170	0.49	8.82	−0.51	1.39	1.27	0.91	0.86	0.47
3	123	1.06	13.88	−0.75	2.26	2.08	1.31	1.24	0.63
4	123	1.28	15.60	−0.74	2.78	2.53	1.62	1.51	0.51
5	123	1.87	17.89	−1.04	3.44	3.13	1.84	1.70	0.47
6	123	2.12	18.98	−1.08	3.93	3.54	1.98	1.85	0.30
7	122	2.47	21.00	−1.35	4.44	3.98	2.27	2.12	0.16
8	124	2.73	16.51	−1.35	4.68	4.13	2.23	1.98	0.13
9	122	3.09	23.37	−1.66	5.38	4.78	2.51	2.37	0.36
10	121	3.28	24.57	−1.89	5.58	4.89	2.61	2.41	0.54
11	39	0.45	4.06	−0.71	1.25	1.70	0.64	0.73	0.13
12	39	0.50	5.36	−0.78	1.46	1.93	0.87	0.91	0.11
13	36	0.72	5.89	−0.68	1.53	1.92	0.81	0.89	0.07
14	40	0.68	5.14	−0.86	1.59	2.06	0.88	0.93	0.04
15	37	0.86	6.60	−0.56	1.82	2.17	0.84	0.90	0.03
16	42	0.78	5.20	−0.72	1.61	2.25	0.93	0.97	0.04
17	38	1.06	7.18	−0.75	1.99	2.30	0.93	0.98	0.07
18	39	0.91	5.78	−0.61	1.69	2.43	0.98	1.04	0.14
19	38	1.22	7.75	−0.75	2.20	2.54	1.05	1.08	0.15
20	40	1.09	6.51	−0.47	1.91	2.61	0.87	0.87	0.13
21	37	1.48	8.38	−0.75	2.36	2.69	1.06	1.08	0.25
22	29	1.89	7.90	−0.31	2.53	2.72	1.37	1.46	0.30

Table 3e: Convenience Prices for No 2 Heating Oil (1967 ¢/gallon).
Average number of contracts/month 9.5.

Maturity (Months)	# Obs	Mean	Max	Min	Std. Dev.	I	II	III	IV
			Convenience Prices			Standard Error of Model Predictions			
Spot	52	33.48	39.43	28.36	2.58				
1	52	0.45	4.66	-0.90	0.98	1.02	0.99	0.98	0.70
2	52	0.77	7.83	-1.53	1.52	1.58	1.54	1.53	0.94
3	52	0.97	8.63	-2.04	1.72	1.79	1.73	1.77	0.85
4	52	1.15	9.05	-2.31	1.84	1.92	1.85	1.93	0.67
5	52	1.33	9.01	-2.52	1.90	1.98	1.91	2.06	0.46
6	48	1.45	9.23	-2.62	2.02	2.08	2.01	2.27	0.29
7	47	1.66	9.15	-2.44	1.97	2.05	1.96	2.25	0.29
8	40	1.86	9.19	-2.03	1.86	1.96	1.88	2.22	0.62
9	33	1.94	9.33	-1.40	1.84	1.93	1.88	2.38	0.90
10	25	1.96	9.37	-1.69	2.09	2.17	1.99	2.93	1.10
11	17	1.29	3.47	-1.94	1.47	1.71	1.23	3.49	1.11
12	10	1.89	4.42	0.10	1.44	1.57	0.97	3.27	0.73
13	6	1.77	4.47	0.34	1.43	1.58	1.37	3.68	0.51

be an increasing function. Moreover, there is considerable instability in the estimated coefficients across the subperiods. We shall say more about this below.

Models II and III offer little improvement over Model I for any of the commercial commodities except copper; for this commodity the Model III parameters are quite stable across subperiods and the standard error of the price predictions is roughly halved relative to Model I. The standard error is around 2 1/2% of the mean spot price and shows no tendency to increase with maturity. Taking account of the possible non-synchronicity of price observations, these errors seem quite small and suggest that with this form of convenience yield function the Brennan-Schwartz model may offer a reasonable approach to the valuation of copper mines.

(ii) The Autonomous Convenience Yield Model

The parameter estimates for Model IV are given in Table 4. For all of the commodities except platinum, and silver for one subperiod, the estimated adjustment coefficient, α, is positive and significantly different from zero. For all of the commodities, m, the long run mean to which the convenience yield reverts is insignificantly different from zero, as is λ^*, the risk adjustment parameter. Inspection of equations (15) and (18) reveals that, given α and the time series of convenience yields, the estimate of m is determined solely from this time series, while the estimate of $\hat{m} = (m - \lambda^*)/\alpha$ is chosen to minimize the price prediction errors. That is, while the actual time series of convenience yields reverts to the long run mean, m, the convenience prices are as if convenience yield rate were expected to revert to $\hat{m}$ and investors were risk neutral. Interestingly, although the estimate of m is predominantly negative, the estimates of $\hat{m}$ are positive for all commodities.

Table 3 also reports the standard errors of the price predictions from Model IV. For all commodities the standard errors are very small relative to the mean commodity price; for gold, platinum and copper the standard errors are generally less than 1% of the mean spot price, for silver and plywood 1–2%, and for heating oil and lumber 2 1/2%. Moreover, there is no significant tendency for the standard errors to increase with maturity. However, it should be recognized that the estimated standard errors are subject to a downward bias on account of the small number of futures price observations per period, and therefore per estimated convenience yield. An approximate correction can be obtained by multiplying the standard errors by $n/(n-1)$ where n is the average number of contracts outstanding per period. n ranges from about 5 for platinum and lumber to almost 11 for gold. Finally, it is interesting to observe that the largest prediction errors are frequently for the very shortest maturity contracts: this suggests that it would not be possible to obtain accurate estimates of the instantaneous convenience yield simply by looking at the very shortest term convenience price.

Table 3f: Convenience Prices for Plywood (1967 \$/thousand sq.feet). Average number of contracts/month 5.8.

| Maturity (Months) | # Obs | Convenience Prices | | | | Standard Error of Model Predictions | | | |
		Mean	Max	Min	Std. Dev.	I	II	III	IV
Spot	46	84.52	118.66	52.47	19.44				
2	46	0.07	7.50	-9.51	2.37	2.40	2.25	2.28	1.53
4	46	0.36	10.28	-10.22	3.61	3.62	3.26	3.32	1.62
6	46	0.76	10.91	-10.50	4.29	4.26	3.68	3.75	1.22
8	46	1.28	13.71	-10.42	4.99	4.85	3.95	4.05	0.85
10	42	2.36	16.33	-9.10	5.21	4.88	4.09	4.06	0.61
12	29	5.10	17.71	-2.68	5.12	4.81	4.62	4.50	1.55
14	8	4.93	12.68	-3.08	4.86	4.44	3.60	4.13	0.84
16	4	6.06	10.06	-0.42	4.05	3.46	2.64	3.37	0.69

Table 3g: Convenience Prices for Lumber (1967 \$/thousand board feet). Average number of contracts/month 5.2.

| Maturity (Months) | # Obs | Convenience Prices | | | | Standard Error of Model Predictions | | | |
		Mean	Max	Min	Std. Dev.	I	II	III	IV
Spot	78	84.87	139.58	38.59	26.23				
2	78	0.22	21.79	−9.11	6.15	6.09	4.71	4.61	2.34
4	78	1.07	32.89	−13.71	9.83	9.70	6.75	6.56	2.14
6	78	1.66	36.99	−15.06	11.83	11.62	7.26	7.03	1.05
8	71	2.23	40.23	−14.60	13.33	13.04	7.56	7.30	1.20
10	48	−0.77	32.18	−14.38	12.21	12.05	7.52	7.33	1.72
12	29	−7.33	10.71	−15.58	5.12	9.98	3.99	4.06	1.72
14	22	−6.73	11.18	−16.69	5.95	10.08	4.85	4.74	4.75

(iii) Speculative Periods and the Probability of Default

As we have already noted, estimation of Model IV required joint estimation of the instantaneous rate of convenience yield each month and the model parameters. The vectors of estimated instantaneous rates of convenience yield are plotted in Figure 1. In this section we are concerned with the precious metals.

For gold the convenience yield fluctuates within a band of 20 cents per ounce per month around zero until the latter half of 1979 when it becomes strongly negative, reaching a minimum of $-$1.62 in February 1980 and slowly reverting to zero over the next twelve months.[19] Since the storage costs of an ounce of gold excluding interest are of the order of 6 cents per ounce per month, the large rewards to storage implied by the futures prices present an apparent profit opportunity. The estimated convenience yield series for silver also shows a massive disturbance in 1979/80. This is related to the large silver purchases of the Hunt brothers in the last quarter of 1979 and their subsequent default; in December 1979, they and others unexpectedly took delivery of 23 million ounces on maturing contracts, driving our estimated convenience yield to over $1 per ounce per month (not shown on the graph). With the collapse of the Hunt's position in March 1980 the estimated convenience yield fell abruptly to $-$72 cents per month, and remained strongly negative for the balance of the year, offering another apparent profit opportunity to potential storers. In February 1983 the convenience yield for silver again fell briefly to $-$19 cents per month; according to *Business Week* the spot price had climbed unexpectedly "raising the specter of new market manipulation and another calamity".[20] In platinum also relatively large negative convenience yield estimates are apparent in 1980 and again in February 1983.

It seems probable that these apparent profit opportunities were due to the very real risk of default in these markets occasioned by the Hunt brothers incident, so that for this reason the convenience yield estimates for this period are biased. Therefore the simple Model I was re-estimated for the three precious metals excluding data drawn from the period September 1979–May 1981 which appeared from Figure 1 to be affected by the Hunt disturbance. The estimated parameters reported in Table 2 correspond to annual rates of convenience yield for gold and silver of 0.17% per year and for platinum 1.9%.

---

[19] In February 1983 it drops briefly again to $-$0.93.

[20] *Business Week* February 28, 1983, p. 31.

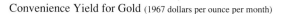

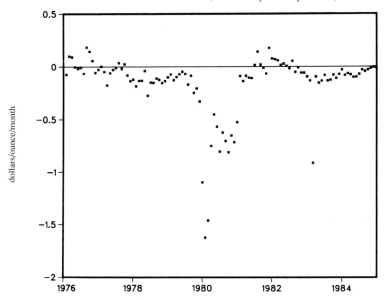

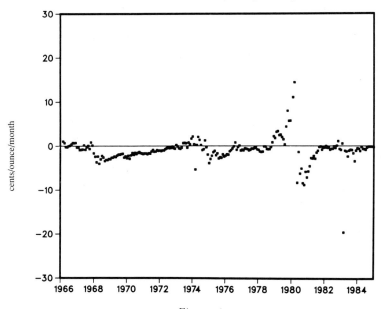

Figure 1.

Convenience Yield for Platinum (1967 dollars per ounce per month)

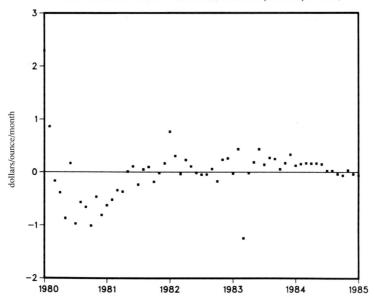

Convenience Yield for Copper (1967 cents per pound per month)

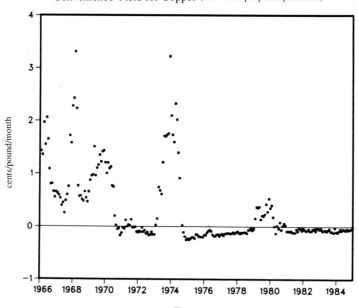

Figure 1.

Convenience Yield for Heating Oil (1967 cents per gallon per month)

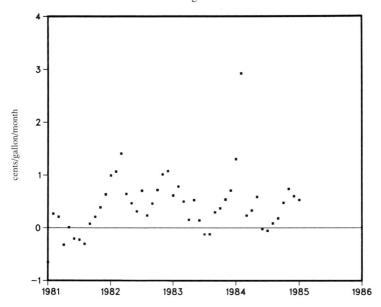

Convenience Yield for Lumber (1967 dollars per thou.bd.ft. per month)

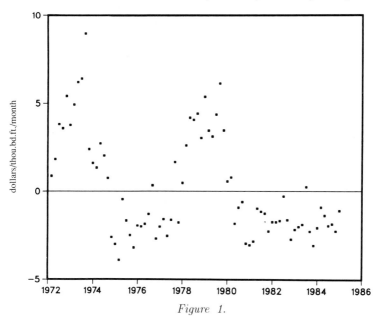

Figure 1.

Convenience Yield for Plywood (1967 dollars per thou.sq.feet per month)

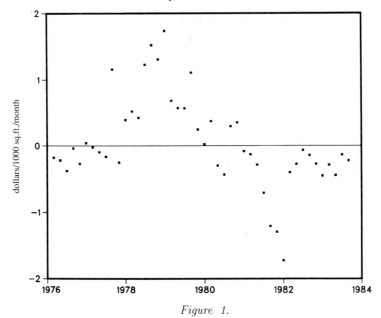

Figure 1.

The standard errors of the Model I convenience price predictions for the precious metals excluding the period of turbulence are also reported in Table 3. For gold the estimated standard errors are now of the order of 1% of the mean spot price, and for platinum and silver of the order of 2% of the mean spot price. This evidence suggests that the simple Brennan-Schwartz model may be useful for valuing claims whose payoffs depend upon the prices of these precious metals.

(iv) Convenience Yields and Inventories

We have argued that the variation of our simple convenience yield estimates is attributable, at least in part, to variations in default risk. This raises the possibility that the true convenience yield for all commodities is constant (and possibly zero) in which case the simple Brennan-Schwartz model will be adequate for pricing all commodity price contingent claims. In this section we test the null hypothesis that the convenience yield is zero against the Kaldor-Working hypothesis that it is negatively related to the level of inventories. Since there exists no formal model of the dependence of the convenience yield on the level of inventories, we have included as

Convenience Yield for Copper (versus inventory/Sales Ratio)

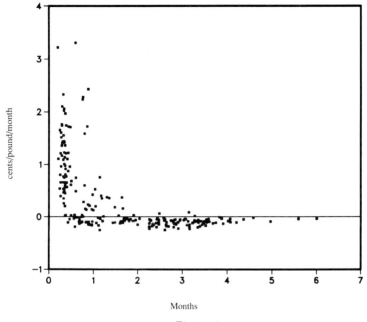

Months

Figure 2.

Appendix B a very simple model under certainty; in this model inventories are demanded because of the fixed reorder cost savings they make possible.

In this section we estimate the relation between our Model IV estimate of the instantaneous rate of convenience yield and the level of inventories. This constitutes a test of the Kaldor-Working theory which implicitly underlies the valuation models presented here and by Brennan and Schwartz (1985).

For each commodity the following non-linear regression is estimated:

$$C = a + b(I/X)^n , \qquad (20)$$

where I is the level of inventories and X a measure of the sales rate.[21] We expect the gross convenience yield to be high for very low inventory levels and to approach zero for high levels. If the costs of storage are constant this implies that $a < 0$, $b > 0$, $n < 0$. On the other hand, if there are

---

[21] A full description of the data can be found in Appendix C.

Table 4: Parameters of the autonomous convenience yield model. $dC = \alpha(m - C)dt + \eta dx_c$. $dS = \mu Sdt + \sigma Sdx_s$. $\rho(C.S)dt = dx_sdx_c$; $\hat{m} = m - \lambda^*/\alpha$; ρ_C and ρ_S serial correlations. Standard errors in parentheses.

	α	m	λ^*	$\hat{m}$	σ	η	$\rho(C.S)$	ρ_C	ρ_S
Gold ($/ounce)									
January 1976–	0.031	-0.018	-0.006	0.176	0.081	0.18	-0.36	-0.03	0.07
December 1984	(0.004)	(0.507)	(0.016)						
January 1976–	0.045	-0.768	-0.035	0.010	0.089	0.20	-0.71	0.33	0.08
June 1980	(0.003)	(0.439)	(0.020)						
July 1980–	0.019	-0.165	-0.011	0.414	0.068	0.16	0.11	-0.40	-0.07
December 1984	(0.004)	(0.675)	(0.013)						
Silver (¢/ounce)									
February 1966–	0.336	-0.915	-0.321	0.040	0.116	10.06	0.52	-0.46	0.08
December 1984	(0.006)	(0.703)	(0.263)						
February 1966–	0.001	-1.889	-0.031	29.11	0.079	0.303	-0.21	-0.17	-0.00
June 1975	(0.006)	(13.09)	(0.015)						
July 1975–	0.332	-1.102	-0.426	0.181	0.144	14.07	0.61	-0.46	0.10
December 1984	(0.009)	(1.121)	(0.373)						

Table 4: (Continued) Parameters of the autonomous convenience yield model. $dC = \alpha(m - C)dt + \eta dx_c$. $dS = \mu S dt + \sigma S dx_s$. $\rho(C,S)dt = dx_s dx_c$; $\hat{m} = m - \lambda^*/\alpha$; ρ_C and ρ_S serial correlations. Standard errors in parentheses.

	α	m	λ^*	$\hat{m}$	σ	η	$\rho(C,S)$	ρ_C	ρ_S
Platinum ($/ounce)									
December 1979–	.000	-56.75	-0.023	u	0.115	0.464	0.19	-0.31	-0.18
December 1984	(.000)	(350.58)	(0.040)						
January 1966–	0.042	0.167	-0.001	0.191	0.181	0.258	0.59	0.02	0.10
December 1984	(0.002)	(0.406)	(0.017)						
January 1966–	0.098	0.631	0.039	0.233	0.087	0.440	0.68	0.04	0.23
June 1975	(0.005)	(0.427)	(0.043)						
July 1975–	0.017	-0.008	-0.006	0.344	0.075	0.058	0.68	-0.19	-0.07
December 1984	(0.002)	(0.269)	(0.005)						
No. 2 Heating Oil (¢/gallon)									
September 1980–	0.262	0.358	0.061	0.125	0.065	0.492	0.59	-0.22	-0.05
December 1984	(0.039)	(0.339)	(0.091)						
Plywood ($/thou.sq.ft)									
January 1976–	0.056	0.189	-0.015	0.459	0.073	0.543	0.58	-0.34	-0.08
August 1983	(0.021)	(0.892)	(0.052)						
Lumber ($/thou.bd.ft.)									
January 1972–	0.051	-0.402	-0.113	0.346	0.116	1.634	0.55	0.04	-0.07
December 1984	(0.016)	(0.728)	(0.113)						

u = undefined

Table 5: Convenience yields and the level of inventories (*t*-ratios in parentheses). $C = a + b(I/X)^n$.

	# Obs.	n	a	b	R^2	Excluding September 1979–May 1981 # Obs.	n	a	b	R^2
Gold (Quarterly) ($/ounce/month)										
1976-I – 1984-IV	36	2.28	-0.01 (0.15)	-0.0005 (3.42)	0.26	29	-4.94	-0.06 (4.40)	7.55 (3.13)	0.27
Silver (Quarterly) (¢/ounce/month)										
1973-I – 1984-IV	48	-0.10	-52.96 (0.47)	67.69 (0.47)	0.00	41	-2.39	-1.46 (4.42)	285.82 (3.53)	0.24
Platinum (Quarterly) ($/ounce/month)										
1979-I – 1984-IV	21	34.00	-0.13 (-1.49)	126x10³ (5.84)	0.64	15	1.00	0.28 (0.59)	-0.03 (0.50)	0.02
						Seasonally Adjusted				
Copper (Monthly) (¢/pound/month)										
January 1966– December 1984	228	-0.89	-0.28 (5.42)	0.49 (15.21)	0.51	228	-0.95	-0.20 (1.72)	0.46 (15.36)	0.53
No.2 Heating Oil (Monthly) (¢/gallon/month)										
September 1980– December 1984	49	6.39	0.53 (5.18)	-0.002 (3.06)	0.17	49	4.30	0.46 (2.16)	-0.02 (3.52)	0.39
Lumber (Bimonthly) ($/1000 bd.ft/month)										
January 1972– December 1984	77	-2.79	3.07 (8.21)	14.01 (10.77)	0.61	77	2.65	-2.86 (5.00)	14.36 (10.93)	0.65
Plywood (Bimonthly) ($/thou.sq.ft./month)										
January 1976– August 1983	46	-2.20	-1.24 (5.30)	4.51 (5.82)	0.43	46	-1.74	-1.74 (5.29)	5.25 (6.47)	0.54

C: convenience yield; *I*: inventories; *X*: sales.

increasing marginal costs of storage the net convenience yield will be large and negative at high inventory levels which would imply that $b < 0$, $n > 0$.

The regression (20) is non-linear in n. It was estimated by repeated ordinary least squares until the optimal value of n was found, and the results are reported in Table 5. When the speculative period is excluded, the estimated relation for both gold and silver is consistent with the Kaldor-Working hypothesis for a commodity with constant marginal costs of storage; the estimated convenience yield function is large and positive at low inventory levels and small and negative at high inventory levels. For platinum, a significant relation between the convenience yield and the level of inventories is found only if the speculative period is included: the reason for this appears to be that it was only during the speculative period that a very low level of inventories was observed: as predicted by the Kaldor-Working theory, this was accompanied by a high rate of convenience yield.

It is at first sight surprising to find evidence of a stochastic convenience yield for these precious metals since, for gold and silver at least, huge inventories are held around the world for non-commercial purposes. However, it should be noted that, once the speculative period is removed, the range of variation in the convenience yield is extremely modest. It appears that this variation, although modest, depends systematically on the level of commercial inventories,[22] presumably because the non-commercial inventories cannot be instantaneously and costlessly 'mined' to replenish the commercial inventories. The lags and costs involved in replenishing the commercial inventories determine the maximum level to which the convenience yield can rise; as we have already noted, a lower bound is determined by the cost of storage which is small for these commodities.

Turning to the commercial commodities, there is a strong negative relation between the net convenience yield for copper and the level of inventories; this is depicted in Figure 2, and is entirely consistent with the predictions of the Kaldor-Working hypothesis when there is a constant cost of storage: the net convenience yield is slightly negative and independent of the level of inventories except at very low inventory levels when the convenience yield rises abruptly.

---

[22] Our data for inventories of the precious metals include only commercial investors.

In the case of heating oil both the level of inventories and the convenience yield estimates exhibited a clear seasonal pattern.[23] To account for this, equation (20) was estimated with and without seasonal adjustment for the commercial commodities.[24] The estimated convenience yield functions for lumber and plywood are similar to that we obtained for copper. However, the relation between the estimated net convenience yield and the level of inventories for heating oil is decreasing and concave. This is consistent with the hypothesis that storage costs for this commodity are a convex function of the level of inventories which is not implausible in view of the specialized facilities required for the storage of this commodity.

In summary, of the seven commodities for which we have data, six show strong evidence that the net rate of convenience yield depends upon the level of inventories as predicted by the Kaldor-Working hypothesis: the seventh commodity, platinum, shows less strong evidence of a similar relation. For six of the commodities the relation between the convenience yield and the level of inventories is consistent with constant marginal costs of storage; for the seventh, heating oil, the relation is consistent with increasing convex storage costs. Notwithstanding the foregoing, for the precious metals there is only modest variation in the estimated convenience yield outside the speculative period.

7. Summary and conclusion

Our primary concern in this paper has been to determine whether valuation models of the type proposed by Brennan and Schwartz (1985) are likely to be useful in evaluating capital projects. Our initial results, using data drawn from the futures markets, were unfavourable: the simple assumption that the convenience yield could be written solely as a function of the spot price was found to result in large pricing errors for most commodities. On closer examination it seems that the Brennan-Schwartz assumption works well for the precious metals except during the period influenced by the Hunt brothers' intervention in the silver market and its aftermath, when fears of default made futures prices unreliable indicators of the convenience yield. For copper also, a version of the Brennan-Schwartz convenience yield function (Model III) worked quite well. However, it was improved on by

---

[23] Note that the specification of Model IV from which the convenience yield estimates were derived is inconsistent with a seasonal in the convenience yield.

[24] Seasonal adjustment was made by introducing bimonthly seasonal dummies into the regression (20).

the autonomous convenience yield model which also performed best for the other commodities.

Estimation of Model IV yielded a time series of estimated rates of convenience yield. The Kaldor-Working theory of the convenience yield predicts that it will be negatively related to the level of inventories. The theory was tested by estimating a non-linear regression of convenience yields on the level of inventories normalized by a measure of sales. For all commodities the convenience yield was negatively related to the level of inventories and for all commodities except heating oil the estimated net convenience yield function was consistent with constant marginal costs of storage; for heating oil an increasing convex marginal cost of storage was indicated.

In related work Gibson and Schwartz (1990) have found that a constant convenience yield assumption does not work well for pricing oil indexed bonds. In subsequent work (Gibson and Schwartz (1989) they have used an approximation to Model IV to compute convenience prices for crude oil.

Appendix A: Overidentification of model IV

Let $Q_{t,\tau}$ denote the convenience price for maturity τ at time t. Then the empirical model to be estimated is

$$C(t) = m\left(1 - e^{-\alpha}\right) + e^{-\alpha}C(t-1) + \xi_t, \qquad (A1)$$

$$Q_{t,\tau} = (m - \lambda^*/\alpha)\left(r - \left(\frac{1 - e^{-\alpha\tau}}{\alpha}\right)\right) + \left(\frac{1 - e^{-\alpha\tau}}{\alpha}\right)C(t) + \epsilon_{t,\tau}, \qquad (A2)$$

$$t = 1,\ldots,T; \quad \tau = 1,\ldots,T.$$

(A1) is the discrete version of the stochastic process for the convenience yield (equation (17) in the text). (A2) is equation (15) with the addition of the model error $\epsilon_{t,\tau}$ which is assumed to be normally distributed. If the model errors were independent, (A2) would be equivalent to the classical factor analysis model with a single factor corresponding to the unobservable convenience yield $C(t)$, and constraints on the factor loadings. However, if the model errors $\epsilon_{t,\tau}$ are correlated the factor model is not identified. We assume that the model errors are contemporaneously but not serially correlated so that

$$E[\epsilon_{t,\tau}, \epsilon_{t,\tau'}] = \omega_{\tau,\tau'}, E[\epsilon_{t,\tau}, \epsilon_{t',\tau'}] = 0 \quad for \ t \neq t'.$$

The model is then identified through the dynamics of the latent convenience yield variable expressed by (A1).[25] Thus note that (A1) implies that $C(t)$ is distributed normally with parameters $(m, \eta^2/(1-e^{-2\alpha}))$, where $\eta^2 = \sigma^2(\xi)$. Then we have the following equations to determine the model parameters:

(i) Contemporaneous Covariances.

Let $S_{\tau\tau'}$ denote the sample covariances between convenience prices of maturities τ, τ'.

Then (A1) and (A2) imply that the parameter estimates should satisfy the $(T^2 + T)/2$ equations

$$S_{\tau\tau'} = \frac{1}{\hat{\alpha}^2}\left(1 - e^{-\hat{\alpha}\tau}\right)\left(1 - e^{-\hat{\alpha}\tau'}\right)\hat{\eta}^2/\left(1 - e^{-2\hat{\alpha}}\right) + w_{\tau\tau'}. \qquad (A3)$$

(ii) Serial Covariances

Let ρ_τ denote the covariance between convenience prices of maturity τ at t and $t+1$. Then (A1) and (A2) imply that the parameter estimates should satisfy the T equations

$$\rho_\tau = \frac{1}{\hat{\alpha}^2}\left(1 - e^{-\hat{\alpha}\tau}\right)^2 e^{2\hat{\alpha}}\hat{\eta}^2/\left(1 - e^{-2\hat{\alpha}}\right). \qquad (A4)$$

(A3) and (A4) are $(T^2 + 3T)/2$ equations for the $(T^2 + T + 4)/2$ parameters $\Omega_{\tau\tau'}$, α, η. Therefore these parameters are identified if $T \geq 2$.

(iii) Cross-section Regressions.

Let a_t, b_t be the estimated coefficients from the (GLS) cross-section regressions

$$Q_{t,\tau} = a_t\tau + b_t\left(1 - e^{-\hat{\alpha}\tau}\right)/\hat{\alpha} \quad (\tau = 1, \ldots, T)\ (t = 1, \ldots, T).$$

Then (A2) implies the following restrictions on the parameter estimates

$$\hat{C}(t) = a_t + b_t, \qquad (A5)$$

$$\hat{m} - \hat{\lambda}^*/\hat{\alpha} = a_t. \qquad (A6)$$

Finally, (A1) implies that

$$\hat{m} = \frac{1}{T}\sum_T C(t). \qquad (A7)$$

---

[25] See Maravall and Aigner (1977) for a model in which identification is achieved by assumptions about the dynamic structure.

(A5) through (A7) comprise $(2T + 1)$ equations, which more than suffice to identify the remaining $(T + 2)$ parameters, $C(t)$, m, λ^*. Hence, as stated in the text, the model is overidentified.

Appendix B: A simple model of the convenience yield

Consider an industry in which a single homogeneous commodity is produced, and in which each of n identical firms faces a known constant sales rate X/n. Let κ denote the cost of holding one unit of the commodity in inventory for one period and let K denote the fixed cost of placing an order for inventory. Then it is well known that, if the interest rate is zero, the optimal order quantity for each firm, q, is given by

$$q = \sqrt{2XK/n\kappa}, \tag{B1}$$

which implies that the average aggregate inventory, $I = nq/2$ is given by

$$I = \sqrt{nXK/2\kappa}. \tag{B2}$$

(B2) may be interpreted as the demand curve for (average) inventory holdings as a function of the periodic holding cost κ. Solving for κ we obtain the inverse demand curve

$$\kappa = \left(\frac{nK}{2}\right)\frac{X}{I^2}. \tag{B3}$$

In (B3) κ is the demand price for an additional unit of inventory, or marginal convenience yield. In the text we employ the marginal net convenience yield, C. This is equal to κ less the physical costs of storing a unit of inventory, $h(I)$. Hence the marginal net convenience yield is given by

$$C = \left(\frac{nK}{2}\right)\frac{X}{I^2} - h(I). \tag{B4}$$

Appendix C: Sales and inventory data

The data series used for the commodity inventories and sales are listed below. All sales figures were converted to an equivalent monthly basis.

Gold: Stocks of gold in the United States held by industry and the futures Exchange. U.S. consumption of gold. U.S. Bureau of Mines, Mineral Industry Surveys, 1976–1984, quarterly.

Silver: Industry, Comex and CBOT stocks. Net industrial consumption. U.S. Bureau of Mines, Mineral Industry Surveys, 1973–1984, quarterly.

Platinum: Stocks held by refiners, importers and dealers in the United States, including depositories of NYME. Platinum sold to consuming industries in the United States. U.S. Bureau of Mines. Mineral Industry Surveys, 1979–1984, quarterly.

Copper: Refined copper stocks in the U.S. Copper deliveries to fabricators in the U.S. International Commodity Yearbook, 1966–1984, monthly.

No. 2 Heating Oil: Stocks of distillate fuel in the U.S. U.S. domestic consumption of distillate fuel oil. International Commodity Yearbook, 1980–1984, monthly.

Lumber: Gross stocks of lumber in the U.S. United States lumber Shipments. International Commodity Yearbook, 1972–1984, monthly.

Plywood: Stocks of softwood plywood in the U.S. Shipments of softwood plywood. U.S. Department of Commerce. Construction Review.

References

Aigner D.J., C. Hsiao, A. Kapteyn, and T. Wansbeck (1983): "Latent Variables Models in Econometrics", in Z. Griliches and M.D. Intriligator (editors), *Handbook of Econometrics*, Vol. II, Amsterdam: North Holland.

Brennan M.J. (1958): "The Supply of Storage", *American Economic Review* 48: 50–72.

Brennan M.J., and E.S. Schwartz (1978): "Finite Difference Methods and Jump Processes Arising in the Pricing of Contingent Claims: A Synthesis", *Journal of Financial and Quantitative Analysis* 13: 461–474.

Brennan M.J., and E.S. Schwartz (1982): "Consistent Regulatory Policy under Uncertainty", *The Bell Journal of Economics* 13: 507–521.

Brennan M.J., and E.S. Schwartz (1985): "Evaluating Natural Resource Investments", *Journal of Business* 58: 135–157.

Cox J.C., J.E. Ingersoll, and S.A. Ross (1981): "The Relation Between Forward Prices and Futures Prices", *Journal of Financial Economics* 9: 321–346.

Dagenais M.G. (1978): "The Use of Incomplete Observations in Multiple Regression Analysis: A Generalized Least Squares Approach", *Journal of Econometrics* 1: 317–328.

Fama E.F., and K.R. French (1987): "Commodity Futures Prices: Some Evidence on Forecast Power, Premiums, and the Theory of Storage", *Journal of Business* 60: 55–74.

Gibson R., and E.S. Schwartz (1989): "Stochastic Convenience Yield and the Pricing of Oil Contingent Claims", UCLA working paper.

Gibson R., and E.S. Schwartz (1990): "Valuation of Long Term Oil-Linked Assets", in this volume.

Jacobs R., and R. Jones (1985): "A Latent Variable Model of the Term Structure of Interest Rates", unpublished manuscript.

Judge C.J., W.E. Griffiths, R.C. Hill, and T.-C. Lee (1980): *The Theory and Practice of Econometrics*, New York: John Wiley & Sons.

Kaldor N. (1939): "Speculation and Economic Stability", *The Review of Economic Studies* 7: 1–27.

Malinvaud E. (1966): *Statistical Methods of Econometrics*, Chicago: Rand McNally & Co.

Maravall A., and D.J. Aigner (1977): "Identification of the Dynamic Shock-Error Model", in D.J. Aigner and A.S. Goldberger (editors), *Latent Variables in Socio-Economic Models*, Amsterdam: North-Holland.

McDonald R., and D. Siegel (1984): "Option Pricing when the Underlying Asset Earns a Below Equilibrium Rate of Return: A Note", *Journal of Finance* 39: 261–266.

Neyman J., and E.L. Scott (1948): "Consistent Estimates of Partially Consistent Observations", *Econometrica* 16: 1–32.

Quandt R.E. (1983): "Computational Problems and Methods", in Z. Griliches and M.D. Intriligator (editors), *Handbook of Econometrics*, Vol. I, Amsterdam: North-Holland.

Telser L.G. (1958): "Futures Trading and the Storage of Cotton and Wheat", *Journal of Political Economy* 66: 233–244.

Working H. (1948): "Theory of the Inverse Carrying Charge in Futures Markets", *Journal of Farm Economics* 30: 1–28.

Working H. (1949): "The Theory of the Price of Storage", *American Economic Review* 39: 1254–1262.

Stochastic Models and Option Values
D. Lund and B. Øksendal (Editors)
© Elsevier Science Publishers B.V. (North-Holland), 1991

Valuation of Long Term Oil-Linked Assets

Rajna Gibson and Eduardo Schwartz

Anderson Graduate School of Management
University of California, Los Angeles, CA 90024-1481

1. Introduction

The purpose of this paper is to present a methodology to price long term real and financial assets whose payoffs are contingent upon the spot price of crude oil. It has recently been stressed in the financial literature that the option pricing framework is more flexible and less restrictive than the traditional net present value approach in valuing natural resources. Indeed, Brennan and Schwartz (1985) derived a contingent claims' valuation approach which rests on the assumption that the spot price of the commodity is a sufficient statistic to value a mine. Their continuous time partial equilibrium valuation model relies on the theory of stochastic optimal control to infer the capital budgeting decisions as well as the value of the ownership's right. Miller and Upton (1985) also make the assumption that the spot price of the commodity is a sufficient statistic but then use a different procedure based on Hotelling's valuation principle to price petroleum properties.

In this paper, our starting point is the same as the one emphasized in the above cited contributions since we also assume that the spot price of oil is at least one of the determinant explanatory variables of any oil-linked real or financial claim. In addition, we argue that the convenience yield of crude oil plays a determinant role in valuing both long and short term oil contingent claims. For that purpose we start by assuming, as Brennan and Schwartz (1985) did, that the convenience yield of oil is a deterministic function of the spot price. We then use the futures markets data in order to extract the forward convenience yields of oil and analyze their time series properties. Our results support the existence of a term structure of forward convenience yields which can be flat, humped, decreasing or increasing. Unfortunately,

This research has been partially supported (Rajna Gibson) by the Swiss National Fund for Scientific Research. We thank Walter Torous, Moon Lee, Steinar Ekern, Marti Subrahmanyam, and Michael Gibson for their helpful comments.

the short maturity of traded futures contracts makes it very doubtful that the up to 10 month ahead estimated forward convenience yields can serve as valid "proxies" of the more distant term convenience yields required to value oil fields or any other long term oil-linked assets.

The second important conjecture we make in this study concerns the specification of the spot price required to value long term contingent claims. Confirming the results Fama and French (1988) obtain in their study of the convenience yields of metals, we have also observed that the prices of oil futures contracts are less volatile than the spot prices and that the former's volatility is a decreasing function of maturity. In particular, Fama and French suggest that this pattern will occur at low inventory levels (high convenience yield levels) and this is precisely what we observed over our reference period (11/26/86–11/18/88). Their empirical tests of the theory of storage are consistent with the fact that supply and demand shocks produce larger fluctuations in near term expected spot prices than in more distant expected spot prices implicitly contained in futures prices.[1] Hence, to price a long term contingent claim, we definitively need to rely on the long term expected spot price of oil given that we want to price such an asset abstracting from short term cyclical shocks affecting the nearest expected spot prices. The latter phenomenon coupled with the fact that there is no true "spot price" for crude oil have induced us to adopt a new methodology which extracts the relevant expected long term or "steady-state" spot price from the futures price.

One of the principal contributions of this paper is thus to show how the existing short term futures and options markets for crude oil can be exploited to extract the forward convenience yields term structure, the "steady-state" spot price of crude oil and the volatility of oil price relatives: three fundamental parameters for the pricing of any real or financial asset contingent upon the expected long term spot price of oil. We then derive a single factor partial equilibrium model whose relevant state variable is the steady-state spot price of oil and apply it to price two oil-linked bonds issued by Standard Oil of Ohio. The analysis of the pricing errors allows us to derive important implications for further research and draws our attention to the fact that the estimation and the specification of the forward convenience yields term structure are essential to any valuation or capital budgeting decision involving crude oil as the underlying physical output.

The paper is organized as follows: Section 2 defines the forward con-

---

[1]   This phenomenon being more pronounced at low inventory levels.

venience yield, the term structure of convenience yields as well as the steady-state spot price of crude oil and analyzes their time series properties relying on futures contracts traded on the NYMEX over the period 11/26/86–11/18/88.

In Section 3, we examine the properties of the crude oil prices' volatility after having extracted the latter from the prices of oil futures options traded on the NYMEX over the period 11/26/86–11/18/88 using Chiras and Manaster's (1978) implied standard deviation weighting scheme.

In Section 4, we describe the two oil-linked bonds issued by Standard Oil of Ohio and emphasize their contingent claims properties.

In Section 5, we derive a single factor partial equilibrium model to value any oil contingent claim and define the boundary conditions under which the partial differential equation will price these two specific assets.

In Section 6, we test the pricing model under four different scenarios concerning the forward convenience yield and the volatility. We then analyze the results and explain the contrasted pricing performance which is observed for the two bonds. This leads us to suggest alternative ways of improving the model's pricing accuracy, in particular by extending the state space to allow for a stochastic convenience yield in addition to the steady state spot price of oil.

Finally, in Section 7, we conclude the paper by suggesting possible research topics that may be worth exploring in order to build a more reliable framework for the pricing of oil deposits, oil leases, oil-linked bonds and any other real or financial asset or decision which is contingent upon the spot price as well as the convenience yield of crude oil.

2. The convenience yield and the spot price of crude oil

Crude oil, as a strategic commercial commodity entitles its owner to quantitative and qualitative services, stemming from the ability to exploit supply shortages, to avoid transportation costs, to control consumption, etc., which do not accrue to the owner of an oil futures contract. This flow of services net of any costs incurred, in particular storage and insurance costs, is defined as the net convenience yield of oil.

In an earlier attempt to price natural resource ownership using the contingent claims valuation approach, Brennan and Schwartz (1985) specified the net convenience yield as a deterministic function of the commodity's spot price alone. However, once we accept the empirical, generally negative, relationship between the level of aggregate inventories and the convenience yield,[2] the former simple specification will hold only if we are willing to admit that the rate of change in crude oil inventories is also a deterministic function of the spot price.[3] Econometric models of oil inventory changes do, however, suggest that the spread between the spot and the official OPEC crude oil prices[4] also determines changes in the level of inventories, a phenomenon which under such a simplifying assumption must then be treated as a random shock occuring during oil markets' turmoils.

Since the purpose of this study is to derive a partial equilibrium framework to value long term real and financial claims on oil, we shall start by restricting ourselves to the Brennan-Schwartz hypothesis which when assuming non stochastic interest rates and a geometric Brownian motion stochastic process for the spot price of oil, enables us to price any futures contract[5] according to the following well known relationship:[6]

$$F_t(S,T) = Se^{(r-\delta)(T-t)}, \tag{1}$$

where:

$F_t(S,t)$ denotes the price at time t of a futures contract expiring at date T.

S denotes the spot price of oil at date t.

r denotes the continuously compounded risk free interest rate

δ denotes the continuously compounded net relative convenience yield.

---

[2] Which is known as the Kaldor-Working hypothesis and for which Fama and French (1988) and Brennan (1986) find strong empirical evidence in the case of agricultural and commercial commodities respectively.

[3] See Brennan (1990).

[4] See Verleger (1982).

[5] For a detailed analysis of (1) and of its derivation, see Brennan and Schwartz (1985) and Ramaswamy and Sundaresan (1985).

[6] Since we abstract from interest rate uncertainty, the prices of futures and otherwise identical forward contracts must be the same. See Cox, Ingersoll and Ross (1981).

The specification and the identification of the convenience yield and of the commodity's spot price are prerequisites for the pricing of any oil contingent claim. This is particularly true for oil claims since oil[7] does not trade for immediate delivery.[8]

For that purpose, we shall present a methodology which exploits the existence of a short term futures market together with some a priori assumptions to infer the convenience yield and the spot price of crude oil. More precisely, we will assume that relationship (1) holds but that in addition, the convenience yield is, like a spot interest rate, a function of the futures contract time to maturity. In other words, we postulate the existence of a term structure of convenience yields and will use pairs of adjacent maturity futures contracts to derive the forward convenience yields implied by that structure. First, we can rewrite relationship (1) as follows:

$$
\begin{aligned}
F_{t_0}(S,T) &= S \exp\left\{(r_T - \delta_T)(T - t_0)\right\} \\
&= S \exp\left\{\sum_{i=1}^{n} \left(r_{t_{i-1},t_i} - \delta_{t_{i-1},t_i}\right)(t_i - t_{i-1})\right\},
\end{aligned}
\tag{2}
$$

where:

δ_T represents the spot convenience yield of a $(T - t_0)$ futures contract at date t_0.

$\delta_{0,1}$ represents the one period $(t_1 - t_0)$ spot convenience yield.

δ_{t_{i-1},t_i} represents the one period $(t_i - t_{i-1})$ forward convenience yield applying (t_{i-1}) periods from now.

r_T represents the $T - t_0$ period spot interest rate.

r_{t_{i-1},t_i} represents the one period $(t_i - t_{i-1})$ forward interest rate applying t_{i-1} periods from t_0.

Moreover, $t_n = T$.

---

[7]   Indeed, even the crude oil contracts traded in the so-called "spot market" are de facto forward contracts with delivery occuring up to 30 days ahead.

[8]   The fact that there is no "true" spot price of oil is a second reason, the first being the theory of storage's implications about the cyclical shocks affecting near term futures contracts' prices — discussed in the introduction — which has led us to "imply" a spot price of crude oil from existing "long term" futures prices.

We can then exploit the fact that futures contracts have a regularly increasing monthly maturity structure to infer the set of forward convenience yields from adjacent maturity futures contracts. Indeed, both the contract maturing at date T and at date $T - 1$ satisfy (2). Hence, dividing the price of the former by that of the latter we obtain the forward one month convenience yield applying $T - 1$ periods ahead:

$$\frac{F(S,T)}{F(S,T-1)} = \exp\left\{(r_{T-1,T} - \delta_{T-1,T})\left(\frac{T-(T-1)}{12}\right)\right\}. \qquad (3)$$

Using annualized interest rates and time periods we can write expression (3) as follows:

$$\delta_{T-1,T} = r_{T-1,T} - 12\ln\left(\frac{F(S,T)}{F(S,T-1)}\right), \qquad (4)$$

where:

$\delta_{T-1,T}$ denotes the $T - 1$ periods ahead annualized 1 month forward convenience yield.

The last step in this procedure consists of extracting the spot price of oil[9] — the relevant state variable to price any long term contingent claim — using the futures prices in conjunction with the estimated forward convenience yields. However, here one must be willing to assume that after the furthest maturity futures contract available, the convenience yield term structure levels off so that the convenience spot and forward yields become the same entity. If we denote by T^* the maturity of the longest available futures contract, we can then invert relationship (1) using the longest $(T^* - 1)$ estimated forward — identified as a spot — convenience yield, hence:

$$\hat{S} = F(T^*)e^{\left(\delta_{T^*-1} - r_{T^*}\right)(T^*-t)}, \qquad (5)$$

where:

---

[9]   By computing the "steady-state" spot price $\hat{S}$ from a longer term futures contract price, we are removing short term temporary variations from the spot price of crude oil as they do not affect the prices of long term assets.

$\delta_{T^*-1} = \delta_{T^*-1,T^*}$ denotes the spot convenience yield assumed to be flat for $T > T*-1$.

$F(T^*)$ denotes the market price of the furthest maturity futures contract at time t.

$\hat{S}$ denotes the steady state spot price of oil at time t.

Applying the procedure given in equations (4) and (5) to real market data, one then becomes able to price any long term security, or any real asset whose payoffs are a function of the spot price of oil and time.

However, before we turn to the development of such a model, we shall present some empirical evidence on the properties of the forward convenience yields and of the steady state spot price. They have both been computed using weekly (every Friday) prices of the futures crude oil contracts traded on the New York Mercantile Exchange and the annualized 6 month T-bill rate during the period 11/26/86 to 11/18/1988.[10]

First of all, as we can see from Table 1,[11] there is a term structure of convenience yields which can be upward or downward sloping as well as humped.[12] This table clearly illustrates the higher volatility of the nearest term forward contracts as well as the fact that the term structure stabilizes itself at the longest maturities. It also shows that negative forward convenience yields, especially in the very short term maturity range, may occur whenever the price of crude oil is low and presumably when the level of inventories is high.

In Figures 1 and 2, we show the evolution of the 4 month ahead and the 6 month ahead forward convenience yields as well as the latter's evolution compared to that of the 7 month futures contract price. These figures as well as Table 2 reporting the mean and the standard deviation of the 2, 4,

---

[10] The data was collected from the Wall Street Journal. More specifically we used settlement prices of the futures contracts.

[11] The dates presented in Table 1 were chosen to emphasize the various patterns characterizing the shape of the term structure of forward convenience yields.

[12] Notice that the forward convenience yields computed using a term structure of interest rates were almost identical and had the same time series properties as the ones computed using a constant interest rate.

Table 1: Term structure of forward convenience yields.

Forward Convenience Yields*	05/12/86	12/24/87	08/19/88
$\delta_{1,2}$	5.63%	12.62%	−1.56%
$\delta_{2,3}$	6.40%	9.65%	−0.23%
$\delta_{3,4}$	9.05%	9.66%	0.09%
$\delta_{4,5}$	8.63%	8.81%	3.00%
$\delta_{5,6}$	8.95%	8.62%	6.18%
$\delta_{6,7}$	8.75%	8.98%	6.32%
West Texas forward contract reference price.	$15.15	$16.65	$15.75

* $\delta_{j,j+1}$ denotes the 1 month ahead annualized forward convenience yield derived from the prices of futures contracts expiring in month j and $j+1$ respectively.

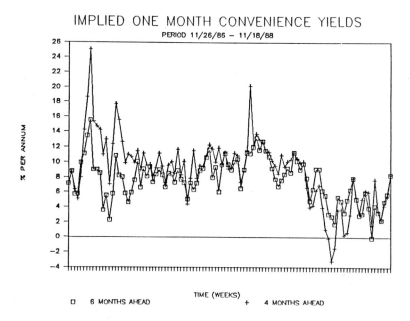

Figure 1.

Table 2: Summary statistics on the forward convenience yields.*

Period 11/26/86–11/18/88	$\delta_{2,t}$	$\delta_{4,t}$	$\delta_{6,t}$	$\delta_{8,t}$
Mean	11.55%	8.96%	7.68%	7.45%
Standard Deviation	7.39%	4.36%	2.81%	3.12%

* δ_j, t denotes the j month ahead ($j = 2, 4, 6, 8$) annualized one month forward convenience yield.

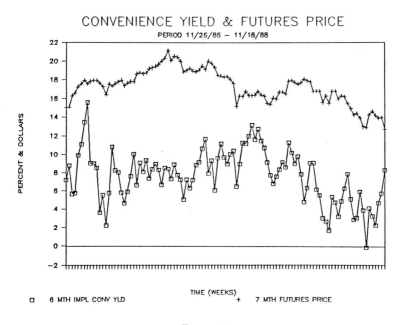

Figure 2.

Figure 2.

6 and 8 month ahead implied 1 month forward convenience yields clearly suggest that over the period analyzed their average level as well as their volatility behaved as decreasing functions of their time to maturity.[13]

---

[13] In the case of crude oil futures contracts, the open interest for each delivery date does not give us additional information about the shape of the term structure of convenience yields since it is systematically lower for distant maturities futures contracts (irrespective of these contracts' price structure) and seems to be primarily related to liquidity considerations. See Duffie (1989).

Table 3: Mean reversion of the forward convenience yields over
the period 11/26/86–11/18/88.

The Regression Model: $\Delta\delta_{j,t} = a + b\delta_{j,t-1} + \epsilon$	$\hat{b}$	t	D-W statistic	R^2	n^*
a) For $\delta_j = \delta_4$:	−0.2274	−3.60	2.12	0.11	103
b) For $\delta_j = \delta_6$:	−0.2875	−4.11	2.05	0.14	103

n^* denotes the number of observations.

OIL FUTURES / STEADY STATE SPOT PRICES
PERIOD 11/26/86 – 11/18/88

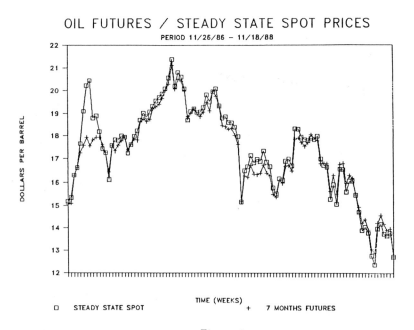

Figure 3.

Table 1 and Table 2 are thus jointly representative of a term structure
of forward convenience yields, a notion which we shall be using extensively
in the remainder of this study, in particular when valuing long term oil
contingent claims and hence requiring relevant proxies of long term forward
convenience yields.

In order to get some insight into the time series behavior of the forward
convenience yields, we ran a series of regressions using the four and the
six month ahead convenience yields which emphasizes their mean reverting

Table 4: Summary statistics on the steady-state spot price and
futures price of crude oil.

| | $\hat{S}$ | F_7 | $\ln(\hat{S}_t/\hat{S}_{t-1})$ | $\ln(F_t/F_{t-1})$ |
	(1)	(2)	(3)	(4)
Mean	$17.41	$17.24	−8.47%	−7.99%
Standard Deviation	$2.00	$1.80	30%	26%

(1) $\hat{S}$ denotes the steady-state spot price of crude oil derived from the 7 month ahead futures
price and the 6 month ahead convenience yield according to the procedure described in
Section 2.
(2) F_7 denotes the settlement price of the seventh closest month maturing futures contract.
(3) and (4) refer to annualized figures obtained over the period 11/26/88–11/18/88.

Table 5: Analysis of the time series behavior of the steady-state
spot price of crude oil.

Regression Model	$\hat{b}$	t	D-W Statistic	R^2	$n*$
a) $\Delta\hat{S}_t = a + b\Delta\hat{S}_{t-1} + \eta_t$	−0.0410	−0.40	1.96	0.00	102
b) $\Delta\hat{S}_t = a' + b'\hat{S}_{t-1} + \epsilon_t$	−0.0429	−1.20	2.00	0.01	103

$n*$ denotes the total number of observations during the period 11/26/86–11/18/88.

pattern. The latter coefficient turned out to be statistically significant for
both δ_4 and δ_6 as can be seen in Table 3.[14]

Although the period of observation may be too short to draw any de-
finitive conclusions[15] about the time-series properties of the forward convenience yields, the empirical evidence presented does suggest the existence
of a mean reverting drift for δ.

As far as the computation of the steady-state spot price of oil is concerned, we had to make a tradeoff between the desired, as long as possible,
maturity of the futures contract and its liquidity. Although futures contracts expiring up to ten months ahead are now available, these contracts
were not traded at the beginning of our reference period; moreover they are
still not actively traded, resulting most of the time in an open interest of

---

[14] Note that for simplicity, we will now write δ_j for $\delta_{j,j+1}$ since it is now clear that we are
always referring ourselves to j month ahead implied 1 month forward convenience yields.
[15] Indeed, over this reference period we only captured a downward rend in oil prices and not
an entire cycle.

less than 1000 contracts. We therefore decided to compute the steady-state spot price using the 7 months ahead futures price in conjunction with δ_6 and equation (5).

Table 4 compares the statistics of the steady-state spot price to those of the 7 month futures price (F_7) over our reference period. Looking at this table as well as at Figure 3, we observe the synchronous evolution of the spot and futures prices as well as the fact that the steady-state spot price and its relative changes are more volatile than those of the futures contracts written on them. In order to gain further insight into the properties of the steady-state spot price, we also performed a series of regressions on $\hat{S}$ that can be seen in Table 5. The results are consistent with the absence of serial correlation in the steady state spot price process and are unable to detect any mean reverting pattern in the spot price's evolution.[16] These results will therefore allow us to use a geometric Brownian motion specification for dS/S in the valuation model since evidence as far as the properties of the drift and of the residuals[17] are concerned tends to consistently support this conjectured process.

Table 6: Analysis of the relationship between the convenience yield and the steady-state spot price of crude oil over the period 11/26/86–11/18/88.

A. OLS Model	$\hat{\beta}$	t	D-W Statistic	$\rho_{S,\delta}(1)$	R^2	n (2)
1. $\delta_{4,t} = \alpha + \beta \hat{S}_t + \mu_t$	0.0113	6.06	0.50	—	0.26	104
2. $\Delta \delta_{4,t} = \alpha' + \beta' \hat{S}_t + \eta_t$	0.0203	5.50	2.44	—	0.23	103
B. Seemingly Unrelated Regression Model (3)						
$\Delta \delta_{4,t} = \alpha'' + \beta'' \delta_{4,t-1} + \varepsilon_t$	−0.1964	−3.50	2.17	0.46	0.11	102

(1) $\rho_{S,\delta}$ denotes the correlation coefficient between the residuals of the $\Delta \delta_{4,t}$ and $\ln(\hat{S}_t/\hat{S}_{t-1})$ processes estimated jointly with the seemingly unrelated regression model.
(2) n denotes the total number of observations.
(3) The seemingly unrelated regression model was fitted to estimated jointly the coefficients of $\Delta \delta_{4,t}$ and $\ln(\hat{S}_t/\hat{S}_{t-1})$ regressing respectively the former variable on $\delta_{4,t-1}$ and the latter on its lagged value. We obtained similar results when restricting the implied spot price process to follow a random walk. Notice that in Part B of the Table R^2 is the coefficient of determination for the entire system.

---

[16] Notice that the same statements were also confirmed for the futures contract price, F_7, over our reference period.

[17] We shall discuss the properties of the variance of the crude oil price process in Section 3.

Finally, we looked at the relationship between the forward convenience yield and the steady-state spot price of oil. In order to eliminate any bias resulting from the fact that $\hat{S}$ was computed using the 6 month ahead convenience yield, this causal relationship was only tested[18] for the four month ahead convenience yield. We summarize the regressions results in Table 6. These results tend to suggest that the deterministic convenience yield hypothesis fails to encompass not only the convenience yield's mean-reverting drift but also the fact that its stochastic process is correlated with the one followed by the spot price of oil.[19,20]

3. The implied standard deviation of crude oil prices

In order to price oil claims, we still need to estimate one additional parameter: the volatility of the relative changes of the spot price of oil. For that purpose, we shall exploit the fact that options on oil futures contracts have been traded on the New York Mercantile Exchange since November 1986 and use their prices to compute the implied spot price volatility. Indeed, if we assume that the futures price is only a function of S and T and if we further allow the spot price of oil to follow a geometric Brownian motion,[21] it can be shown that the futures and the spot price relative changes then have the same volatility.[22] Hence, for all three nearest term futures option contracts available we computed every week the implied standard deviation of both put and calls[23] using Black's European futures written option pricing formula (1976) in conjunction with the market prices of these options. We then computed a weighted, both put and call, implied standard deviation. The weight associated to each

---

[18] Although not reported, the results for others δ_j's are of the same order of magnitude.

[19] Comparing the results of Table 3(a) with those in part B of Table 6, we see that the mean reversion coefficient in $\triangle\delta_{4,t}$ shouldn't be estimated without accounting for the high degree of correlation between the residuals of the implied spot price process $(\ln(\hat{S}_t/\hat{S}_{t-1}))$ and the convenience yield's process $\triangle\delta_{4,t}$. Indeed, the seemingly unrelated regression model used in Table 6 points out that the correlation coefficient between the two processes' residuals is equal to 0.46.

[20] Brennan (1990) modelled the convenience yield as a simple first order mean reverting process and already found that for most commercial commodities this specification outperforms the simple Brennan-Schwartz deterministic specification of the convenience yield.

[21] Note that we also require a non-stochastic specification of the net convenience yield and of the risk-free rate for this relationship to hold.

[22] See, in particular, Ramaswamy and Sundaresan (1985) for the derivation.

[23] We excluded all options whose prices violated obvious arbitrage boundaries and we keep at most two in-the-money and two out-of-the-money calls (and puts) for each distinct maturity to partially deal with the fact that the traded options are of the American type.

individual option's implied volatility was set equal to its price elasticity with respect to volatility according to a procedure described by Chiras and Manaster (1978). Hence, the weighted implied standard deviation, WISD, satisfies the following formula:

$$\text{WISD} = \frac{\sum\limits_{j=1}^{N} \sigma_j \frac{\partial w_j}{\partial \sigma_j} \cdot \frac{\sigma_j}{w_j}}{\sum\limits_{j=1}^{N} \frac{\partial w_j}{\partial \sigma_j} \cdot \frac{\sigma_j}{w_j}}, \tag{6}$$

where:

$\text{WISD} =$ the weighted implied standard deviation of the spot price of oil at a specific observation date.

$w_j =$ price of the jth option at the observation date.

$\sigma_j =$ implied standard deviation of the j-th option — put or call — at the observation date.

$\frac{\partial w_j}{\partial \sigma_j} \cdot \frac{\sigma_j}{w_j} =$ price elasticity of the j-th option with respect to its implied standard deviation.

$N =$ total number of options on a given observation date.

The evolution of the weighted implied standard deviation over our reference period is illustrated in Figure 4. It clearly suggests that the volatility extracted from short term option contracts[24] written on short term futures contracts is far from being constant over time. Also, if we compare the average implied standard deviation of oil prices (33%) to the historical volatilities of $\ln(F_t \mid F_{t-1})$ and $\ln(\hat{S}_t \mid \hat{S}_{t-1})$ presented in Table 4, we observe the well known fact that the shorter the maturity of the options, and hence of their underlying futures contracts, the higher and the more volatile[25]

---

[24] The longest available option maturity is three months and each futures contract has a time to maturity which exceeds at most by twenty days the time to maturity of its associated option contract.

[25] See in particular Fama and French (1988). Their evidence supports the decreasing votality pattern of futures contracts as a function of their maturity. The authors suggest that for commodities this phenomenon is most apparent when inventories are low.

their estimated WISD[26] compared to the historical volatilities of futures and spot relative prices.

To gain some additional insights into the time series properties of the WISD, we examine in Table 7 whether a mean-reverting process and first order serial correlation could be detected over our reference period. Although the R^2 coefficients are fairly low, each regression's explanatory variable is significantly different from zero, suggesting that over the period covered the volatility of oil may have displayed a mean-reverting drift as well as serial correlation. Given the limited improvement of previous stochastic volatility models applied to common stocks options (see Wiggins (1987) and Scott (1987)) and to foreign currency options (see Chesney and Scott (1989)), it might be worthwhile to explore first whether oil volatility bears a simpler deterministic relationship to the level, as in the case of the constant elasticity of variance specification,[27] or to past innovations, as in the case of a first order autoregressive conditional heteroskedastic (ARCH) specification,[28] of the crude oil price process.

---

[26] Notice that the mean WISD over the period was equal to 33% and its standard deviation was equal to 10.85%.

[27] See Cox (1975). In the constant elasticity of variance model, it is assumed that the state variable S follows the diffusion process:

$$dS = \mu S dt + \delta S^{\theta/2} dz \,,$$

where θ denotes the constant elasticity of the variance of relative price changes and its usually expected to be strictly less than 2 when the CEV model holds for common stock prices.

[28] In a first-order ARCH process (see Engle (1982) and Jorion (1987)), the state variable S follows the following stochastic process:

$$\frac{dS}{S} = \mu dt + \sigma(S,t) dt \,,$$

with the conditional variance V_t being explicitly modelled as a linear function of past squared innovations. For a first order ARCH process we can thus write:

$$V_t = a_0 + a_1 \left(\ln \left(\frac{S_{t-1}}{S_{t-2}} \right) - \mu \right)^2 = a_0 + a_1 e_{t-1}^2 \,.$$

When there is no heteroskedasticity in the data we should observe $a_1 = 0$ and $a_0 = \sigma^2 = $ constant.

88   *Rajna Gibson and Eduardo Schwartz*

Table 7: The time series behavior of the implied standard deviation
of crude oil price relatives.

Regression Model	$\hat{\beta}$	t	D-W Statistic	R^2	n^*
1. $\Delta\mathrm{WISD}_t = \alpha + \beta\Delta\mathrm{WISD}_{t-1}+\epsilon_t$	−0.2462	−2.55	2.06	0.06	103
2. $\Delta\mathrm{WISD}_t = \alpha + \beta\mathrm{WISD}_{t-1}+\mu_t$	−0.1316	−2.19	2.28	0.05	104

n* denotes the total number of observations over the period 11/26/86–11/l8/88.

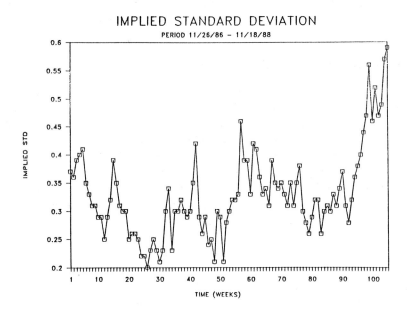

Figure 4.

To test the former hypothesis, we relied on Beckers' (1980) methodology,
using the weekly annualized weighted implied standard deviations (instead
of the absolute value of $\ln(S_{t+1}/S_t)$, the proxy used by the author on daily
data) to run the following regression:[29]

$$\ln \mathrm{WISD}_t = \alpha + \beta \ln S_t + \mu_t$$

---

[29] We should, however, be careful when interpreting these discretely sampled results of an
instantaneous relationship.

The results over our reference period can be summarized as follows:

$\hat{\beta}$	t	R^2	n
-1.36	10.77	0.53	104

They suggest that the volatility of oil price relative changes was negatively related to the oil price level and that the constant elasticity of variance[30] parameter, θ, had an average value of -0.72 which was significantly less than 2, thus suggesting that a CEV model might be better suited than the lognormal one to describe the evolution of the spot price of oil.

4. Description of the oil-linked bonds

As already mentioned, the objective of this paper is to suggest a new methodology to value long term assets whose payoffs are contingent upon the spot price of oil. For that purpose, we have shown how existing oil futures and options markets can be exploited to extract the relevant parameters, namely the convenience yields, the steady-state spot price of oil and the volatility of oil prices, and we have examined their time series properties over a two year reference period.

We shall now illustrate how these parameters can be used in conjunction with a single-factor valuation model to derive the theoretical prices of a specific oil contingent claim. More specifically, we shall price two oil-linked bonds which have been issued by Standard Oil of Ohio Company at the end of June 1986. Each of them represents $37,500,000 face value of zero coupon notes maturing December 15, 1990 and March 15, 1992 respectively. At maturity, the holder of each $1000 face value note is entitled to receive par plus an additional amount equal to the excess, if it exists, of an amount based on a calculated crude oil price for West Texas Intermediate (WTI) crude oil (not exceeding $40) over $25 multiplied by 170 barrels for the 1990 issue and by 200 barrels for the 1992 issue, respectively. In addition, the holder of each 1990 (1992) note can redeem it before maturity at the above

---

[30] See footnote (27). Notice that in the regression:

$$\beta = \frac{\theta - 2}{2}.$$

conditions[31] on the first and fifteenth of each month beginning January 1, 1990 (April 1, 1991). Whether it be at maturity or at an early redemption date, the relevant "settlement price" is defined as the average of the closing prices of the New York Mercantile Exchange light sweet crude oil futures contract for the closest traded month during a predetermined "trading period". The trading period is defined as a one month period ending twenty two days before the relevant redemption or maturity date.

Clearly, each Standard Oil issue can be viewed as a portfolio consisting of an ordinary zero-coupon bond plus one "quasi-American" (long term) call option with an exercise price of $25 held long and one quasi American (long term) call option with an exercise price of $40 held short, both of them written on the same oil futures contract. Notice that it is the bondholders' put option which makes it in fact a "quasi-American" vertical spread since it enables him, during the last year of the bond's life, to exercise his long call and liquidate the short one every two weeks.

In order to price these two bonds, we shall make the following three simplifying assumptions:

(1) Beginning January 1, 1990 and April 1, 1991 respectively, the calls can be exercised at any time (instead of every two weeks).
(2) The underlying security of both options is a fixed maturity (two weeks) futures contract. This simplification is based on the average maturity of the closest futures contract over the "trading period".
(3) The "settlement" price will be based upon the fixed maturity futures contract's price at the redemption or maturity date (instead of the one month average price specified in the prospectus).

5. The oil-linked bond valuation model

We shall value these long term securities indexed on the price of crude oil assuming that there is no default-risk and that interest rates are non-stochastic. We shall also assume that the spot price of oil is the only relevant state variable, in addition to time or time to maturity $\tau = (T - t)$, which determines the price $P(S, \tau)$ of any oil contingent claim and that it follows the following stochastic process:

$$\frac{dS}{S} = \mu dt + \sigma dz \,. \tag{7}$$

---

[31] Notice that "par" is then equivalent to the discounted face value of the issue.

If we further assume that the price of the claim $P(S, \tau)$ is twice continuously differentiable, we can use Ito's lemma to define its instantaneous price change as follows:

$$
\begin{aligned}
dP &= P_s dS - P_\tau dt + \frac{1}{2} P_{ss} (dS)^2 \\
&= \left[\mu S P_s - P_\tau + \frac{1}{2} \sigma^2 S^2 P_{ss} \right] dt + S \sigma P_s dz \,.
\end{aligned}
\tag{8}
$$

In addition, since we have assumed that the convenience yield is constant, we know that the price of any oil futures contract $F(S, T)$ will satisfy relationship (1). Moreover, we know that by Ito's lemma dF must satisfy:

$$
dF = \left(-F_\tau + \frac{1}{2} F_{ss} \sigma^2 S^2 \right) dt + F_s dS \,.
\tag{9}
$$

Hence, by going long in one claim $P(S, \tau)$ and shorting (P_s/F_s) futures contracts maturing at the same date T, it is obvious from equations (8) and (9) that we are de facto holding a riskless position. Under the standard perfect market assumptions, the instantaneous return on such a position must then be equal to rP, i.e., the risk-free rate times the value of the funds invested. Combining the no arbitrage condition with relationships (9), (8), and (1), it can easily be shown that the price of the claim must, in equilibrium, satisfy the following partial differential equation:

$$
\frac{1}{2} \sigma^2 S^2 P_{ss} + (r - \delta) S P_s - P_\tau - rP = 0 \,.
\tag{10}
$$

We know that under our set of assumptions the price of any real or financial claim whose payoffs are only a function of the spot price of oil and time must satisfy the general valuation equation (10). In particular, the oil-linked bonds issued by Standard Oil will also satisfy equation (10) subject to the following boundary conditions:[32]

a) *The initial condition:*

$$
P_i \left(\hat{S}, 0 \right) = FV + \mathrm{Max} \left[0; \mathrm{Min} \left(\hat{S} e^{(r-\delta) \tau^*} - 25, 15 \right) Q_i \right] ,
\tag{11}
$$

where:

---

[32] Notice that these boundary conditions are based on the three simplifying assumptions stated in Section 4 as well as upon the more general conditions described in that same section.

$P_i(\hat{S}, 0)$ denotes the price of the ith bond ($i = 1, 2$) at maturity, given that $i = 1$ refers to the 1990 maturing issue and $i = 2$ to the 1992 maturing issue.

$\hat{S}$ denotes the steady-state spot price of oil.

$\hat{S}e^{(r-\delta)\tau^*}$ denotes the price of a fixed maturity τ^* (equal to 14 days) futures contract deliverable at the maturity or upon early exercise of the bondholder's option.

Q_i denotes the quantity of oil barrels accruing upon exercise of the option at maturity (or upon early redemption) such that $Q_1 = 170$ barrels and $Q_2 = 200$ barrels.

FV denotes the face value (1000\$) of the bond at maturity.

b) *The early exercise condition:*

$$P_i\left(\hat{S}, \tau\right) \geq FVe^{-r(T-t)} + \text{Max}\left[0; \text{Min}\left(\hat{S}e^{(r-\delta)\tau^*} - 25, 15\right)Q_i\right],$$
$$\text{for all } \tau \leq \tau_c,$$

$$(12)$$

where:

$\tau_c = T - t_c$ represents the time to maturity of the bond at its first early redemption date, t_c, which corresponds to January 1, 1990 for $i = 1$ and to April 1, 1991 for $i = 2$.

c) *The lower boundary condition:*[33]

$$P_i(0, \tau) = FVe^{-r(T-t)},$$

$$(13)$$

---

[33] As stated above, the lower boundary condition abstracts from default risk and bankruptcy issues. This arises as a corollary of the original definition of the single factor pricing model (which has been derived assuming no default risk and no interest rate uncertainty). It remains an empirical issue — for high quality bonds — to determine whether the value of the firm should also be introduced as a state variable in the pricing model to account for the issuer's risk and its evolution over time.

d) *The upper boundary condition:*[34]

$$P_i(\infty, \tau) = FVe^{-r(T-t)} + [(40 - 25)Q_i]e^{-r(t_c-t)},$$
$$\forall t < t_c,$$
(14a)

$$P_i(\infty, \tau) = FVe^{-r(T-t)} + [(40 - 25)Q_i],$$
$$\forall t \geq t_c.$$
(14b)

There is no analytical solution to the system formed by the linear partial differential equation (10) subject to the boundary conditions (11) to (14). We therefore solved it numerically using an implicit finite difference method to value the two oil-linked bonds issued by Standard Oil of Ohio Company.

6. The theoretical prices of the oil-linked bonds: Main empirical evidence

Since their issuance, the Standard Oil securities have been infrequently traded, resulting on most observation dates[35] in the absence of transaction prices and in bid ask quotes which sometimes remained unchanged during up to two weeks. Furthermore, the bid and ask spread was most of the time too large, averaging 10% of the bid price, to allow any meaningful test of the robustness of our model based solely on the bid and ask prices (or on the "average price"). Hence, we decided to consider only those observation dates for which there was a transaction available for either of the two issues. This resulted in a very small sample of 28 relevant transaction prices for the Standard Oil 1990 issue and of 29 transaction prices for the Standard Oil 1992 issue. The first transaction has been recorded the 22 August, 1986 and the last one the 27 September, 1988. Unfortunately, we were unable to obtain any information on the volume which has been traded at each

---

[34] Notice that such a condition is required since we solved the partial differential equation (10) numerically. Preliminary tests of the valuation program using this fixed end point boundary gave the same results as those based on the less restrictive boundary condition on the second derivative: $\partial^2 P/\partial S^2 = 0$ as $S \to \infty$ which has finally been used in the final version of the numerical valuation program.

[35] We would like to thank Mr. Milorad Rado of Pictet & Cie., Geneva, Switzerland, for having prvided daily bid and ask quotes as well as the transaction prices of these two bonds over the period 08/01/1986 to 10/14/1988.

of these 57 distinct transactions. In order to apply the valuation model to these bonds, we estimated the steady-state spot price of oil, $\hat{S}$, at each transaction date using equation (5) in conjunction with the market price of the 7 month ahead futures contract, the 6 month[36] ahead annualized one month forward convenience yield and the annualized 6 month T-Bill rate.

We also computed at each transaction date the riskfree interest rate r as being equal to the yield to maturity of a government bond whose maturity is as close as possible to the one of each Standard Oil issue — respectively, the yield of the 13% November, 1990 Treasury Note and of the 14 5/8% February, 1992 Treasury Note — to which we added the relevant monthly average spread between the yield of AA rated ordinary fixed-income securities[37] and the yield of long term riskless bonds in order to account for the issuer's intrinsic quality.

As far as the other parameters necessary to value the oil-linked bonds, namely, the volatility of oil and the forward convenience yield, are concerned, we examined four different scenarios which can be summarized as follows:

Scenario 1: relies at each transaction date on the then prevailing weighted implied standard deviation, computed from the oil futures options' prices using the method described in Section 3, as well as the then prevailing forward one month convenience yield, computed with the methodology described in Section 2, to price these two bonds. Ideally, we would like to use a forward convenience yield whose maturity matches the time to maturity of each bond. However, we have already mentioned the absence of futures contracts which exceed one year to maturity. Moreover, the tradeoff between the liquidity[38] of longer term futures contracts and their maturity led us to choose the 7 month futures contract and, hence, the 6 month ahead forward convenience yield as the proxy for all longer term forward convenience yields or, in other words, to assume that thereafter the term structure of convenience yields is flat. Hence, at each observation

---

[36] We decided to rely on the 6 month ahead implied 1 month forward convenience yield since the longer term futures contracts necessary to obtain more distant forward convenience yields did not trade regularly at the beginning of our reference period and since, later on, even when they traded, their open interest was usually less than 1000 contracts.

[37] This is the rating given by the Standard & Poor's bond guide to the ordinary bonds issued by Standard Oil. The yield spread between AA rated bonds and long term Government Securities has also been computed using the latter guide's data.

[38] Since longer term futures contracts do not trade regularly and didn't at all trade during the period extending from August 1986 to January 1987, we couldn't rely on their prices to get a proxy for the long term forward convenience yield each transaction date.

date t, our first scenario is based on $\delta_{6,t}$ and WISD_t to solve the partial differential equation (10).

Scenario 2: As far as the implied volatility is concerned we used the same assumption as in Scenario 1. However, and in order to account for:

a) the fact that the forward — and hence spot — convenience yields are less volatile the longer their maturity,

b) the mean-reverting pattern in the convenience yields detected in Section 2,

we decided to use the average of the weekly 8 month ahead annualized forward convenience yields $\delta_{8,t}$ over the period 11/26/86 to 11/27/87 to compute the bond values on the transaction dates during this period and the average of the weekly 8 month ahead annualized forward convenience yields over the period 12/04/87 to 11/18/88 to compute the bond values over the remaining transaction dates.

The two average values are respectively equal to 7.10% and 7.70% and were, at each transaction date, combined with WISD_t to value the two bonds using the partial differential equation (10).

Scenario 3: In order to capture the expected — "more stable" — behavior of oil volatility over the long run and in view of the rather weak evidence in favor of a mean reverting drift in oil volatility found in Section 3, we decided to identify the long term oil volatility with a simple arithmetical average of WISD_t over the total period and to use it in conjunction with a varying forward convenience yield $\delta_{6,t}$ — as given in Scenario 1 — to price the bonds. Hence, under this scenario, we value the bonds using $\delta_{6,t}$ and $\text{WISD} = 33\%$ during the whole period.

Scenario 4: Under this scenario, we are combining the stabilizing pattern of both the convenience yield and the volatility over the long run. In order to eliminate the short term variations in these two parameters we value the two bonds at each transaction date using the average volatility (WISD = 33%) over the period as well as the average convenience yield $\delta_8 = 7.10\%$ for the period 11/18/86 to 11/27/87 and $\delta_8 = 7.70\%$ for the transactions occuring thereafter. Hence, scenario 4 relies on:

a) $\text{WISD} = 33\%$ and $\delta_8 = 7.10\%$ until 11/27/87,

b) $\text{WISD} = 33\%$ and $\delta_8 = 7.70\%$ thereafter.

We shall now present in Tables 8 and 9 the results obtained for each bond under the four scenarios just described. We will rely on the mean relative pricing error,[39] on its standard error and on the mean root squared error expressed in relative terms to analyze the model's valuation performance[40] under each set of assumptions about the forward convenience yield and the oil price relatives' volatility.

Looking at Tables 8 and 9, we can immediately observe that under all scenarios the model performs much better in valuing the 1990 Standard Oil issue than the 1992 Standard Oil issue. This suggests that the pricing bias induced by computing the long term forward convenience yield and the expected long term volatility using short term options and futures data increases with the maturity of the contingent claim. In other words, the short term variations in these two parameters and in particular the time series behavior of the short term forward convenience yields used as our estimates of the long term[41] forward convenience yield explain why the performance of a single factor model decreases as the maturity of the claim we are valuing is lengthened. Both the mean relative pricing error as well as the root mean squared error which are consistently larger, almost five and two times respectively under scenario 4, for the 1992 issue confirm this hypothesis.

The second general pattern which emerges from the observed results is that the model systematically underprices both issues and that, looking at the mean relative pricing error, this statement persists under all four

---

[39] The mean relative pricing error e is defined as:

$$e = \frac{1}{n} \sum_{t=1}^{n} \frac{\hat{P}_t - P_t}{P_t},$$

where:

n denotes the total number of transactions for the bond,
P_t denotes the market price of the bond at transactions date t,
$\hat{P}_t$ denotes the theoretical value of the bond at transaction date t.

[40] Given the lack of trading characterizing the oil-linked bonds issued by Standard Oil, one should be aware, when analyzing our results, that even the quoted transaction prices of these issues can't be treated as "fair" market prices. Unfortunately, we have not been able to obtain data for other oil-linked bonds for which we could have abstracted from liquidity and volume related problems when testing the performance of the model.

[41] The empirical evidence in Section 2 supports the existence of a term structure of convenience yields as well as a decreasing pattern in the volatility of the forward convenience yields as a function of their time to maturity.

Table 8: Analysis of the model's pricing performance standard oil 1990 issue.

	Scenario 1[2]	Scenario 2[2]	Scenario 3	Scenario 4	n^1
	$\delta_{6,t}$, WISD$_t$: both vary with t.	$\delta_8 = 7.1\%$ until 11/26/87. $\delta_8 = 7.7\%$ after. WISD$_t$ varies with t.	$\delta_{6,t}$ varies with t. WISD=33%.	$\delta_8 = 7.1\%$ until 11/27/87. $\delta_8 = 7.7\%$ after. WISD=33%.	28
Mean relative pricing error e	−0.0524	−0.0507	−0.0308	−0.0264	
Standard deviation of e	(0.1080)	(0.0771)	(0.0940)	(0.0531)	
Mean root squared error MRE	0.1200	0.0923	0.0989	0.0593	

[1] n denotes the total number of transaction dates
[2] Since options markets only opened in November 1986 for the four transactions preceding that date, we computed the bond's theoretical value using the first available WISD$_t$.

scenarios. However, and this remains true for both Standard Oil issues, the mispricing tends to decrease if we abstract from the intertemporal variation in either of the two crucial parameters. First of all, we can see by comparing scenario 3 with scenario 1 that if we use the average volatility (WISD = 33%) over the whole period, we are able to reduce the relative mean pricing error as well as the root mean square error by 2% (in absolute terms) for both bonds. This clearly suggests that the short term cyclical movements in oil volatility — captured with scenario 1 — must be eliminated in order to price long term oil-linked securities and that updating the implied standard deviations, computed from short term oil futures options prices, defines an inadequate way of capturing the expected trend in oil volatility over the long run. As can be seen by looking at scenarios 3 and 4, it remains, in the absence of a correct specification of the stochastic (or deterministic) process of oil volatility, more appropriate to use the constant volatility estimate since it is certainly more consistent with the slight mean reverting pattern detected in this parameter's evolution.

The third fundamental observation is concerned with the relevance of

Table 9: Analysis of the model's pricing performance standard oil 1992 issue.

	Scenario 1[2]	Scenario 2[2]	Scenario 3	Scenario 4	n[1]
	$\delta_{6,t}$, WISD_t: both vary with t.	$\delta_8 = 7.1\%$ until 11/26/87. $\delta_8 = 7.7\%$ after. WISD_t varies with t.	$\delta_{6,t}$ varies with t. WISD=33%.	$\delta_8 = 7.1\%$ until 11/27/87. $\delta_8 = 7.7\%$ after. WISD=33%.	29
Mean relative pricing error e	−0.1184	−0.1177	−0.1023	−0.1015	
Standard deviation of e	(0.1326)	(0.0989)	(0.1119)	(0.0737)	
Mean root squared error MRE	0.1778	0.1537	0.1516	0.1254	

[1] n denotes the total number of transaction dates
[2] Since option markets only opened in November 1986, for the 4 transaction dates that preceding that date, we computed the bond's theoretical value using the first available WISD_t.

the two proxies of the long term convenience yield used in this study. By looking at the results of scenario 1 in both tables, we can state that the current level of the 6 month ahead forward convenience yield is not a proper estimate of the — between two to five years ahead — forward convenience yields required to value these bonds. This could be explained by the fact that the term structure of convenience yields flattens out at a more distant maturity than 6 months — as we had to assume given the absence of longer term futures contracts traded on a regular basis — as well as by the fact that the volatility of the 6 months ahead forward convenience yield is too high to capture the more static behavior of longer term forward convenience yields. Moreover, scenario 2 in which we tried to solve these two problems using a constant average convenience yield — whose maturity was longer (eight months instead of six) than in scenario 1 — for each distinct yearly period,[42] only slightly improved the results for both bonds. The magnitude

---

[42] We also valued both bonds using the average of the 6 month ahead forward convenience yield over the total period ($\delta_6 = 7.68\%$) yet the results were of the same order of magnitude as those obtained with scenario 2.

of the underpricing remained almost the same and only the volatility of the mispricing, by looking at the standard error of e and at MRE, has effectively been reduced. Also, when we compare the results for both bonds, we can see that an 8 month ahead forward yield can price the shorter issue better than the 1992 issue. This fact confirms that the term structure of convenience yields levels off at more distant maturities than the ones available from existing futures markets data.

Finally, we can say that while the constant convenience yield assumption, in scenario 2, slightly improved the results and is in this respect consistent with the mean-reverting pattern of the forward convenience yields observed in Section 2, the reference period used was too short to lead to a proper estimate of the expected value of δ over the long run. The latter fact and the slow levelling off of the term structure of convenience yields clearly suggest that one can only improve the results by:

(1) Relying on a longer time series of futures price data encompassing a whole cycle of oil price variations.
(2) Determining the stochastic process which drives the instantaneous convenience yield.

7. Implications for further research

The results presented in this study have important implications for the pricing of long term real and financial assets whose payoffs are contingent upon the spot price of oil. First of all, we argue in light of our empirical results that the spot price of oil is not a sufficient statistic to accurately value any type of oil security or resources.[43] Instead, we find that the convenience yield is a fundamental parameter whose level and stochastic evolution considerably influence the theoretical value of these assets. In particular, the failure to recognize the higher level and volatility of the nearest term convenience yields may lead us to substantially overprice short term oil-linked securities. On the other hand, failure to properly specify the levelling off of the convenience yields' term structure can lead us to underprice long term oil-linked assets. The need to properly specify the stochastic process driving the instantaneous convenience yield thus seems to be the first step in order to build a two factor model — based on the spot price of oil and on the latter variable — whose performance could then be compared to that of the single factor contingent claim valuation

---

[43] In particular, see Miller and Upton (1985) and Brennan and Schwartz (1985).

model presented in this study. Finally, it may be worthwhile to exploit the
conclusions of this study in order to extend them for the purpose of valuing
real and financial claims written on other commodities whose convenience
yield is an important parameter and hence needs to be explicitly considered
and modelled in any meaningful valuation model derived.

References

Beckers S. (1980): "The Constant Elasticity of Variance Model and its
    Implications for Option Pricing", *Journal of Finance* 35(3): 661–673.

Bjerksund P., and S. Ekern (1989): "Managing Investment Opportunities
    Under Price Uncertainty: From 'Last Chance' to 'Wait and See'
    Strategies", unpublished paper, Norwegian School of Economics and
    Business Administration, Bergen, September.

Black F. (1976): "The Pricing of Commodities Contracts", *Journal of
    Financial Economics* 3: 167–178.

Brennan M. (1990): "The Price of Convenience and the Valuation of
    Commodity Contingent Claims", in this volume.

Brennan M., and E. Schwartz (1985): "Evaluating Natural Resource In-
    vestments", *Journal of Business* 58(2): 135–157.

Chesney M., and L. Scott (1989): "An Empirical Analysis of European
    Currency Options: A Comparison of the Modified Black-Scholes Model
    and a Random Variance Model", *Journal of Financial and Quantita-
    tive Analysis*, forthcoming.

Chiras D., and S. Manaster (1978): "The Information Content of Option
    Prices and a Test of Market Efficiency", *Journal of Financial Eco-
    nomics* 6: 213–234.

Cox J. (1975): "Notes on Option Pricing 1: Constant Elasticity of Variance
    Diffusions", Working Paper, Stanford University.

Cox J., J. Ingersoll, and S. Ross (1981): "The Relation Between Forward
    and Futures Prices", *Journal of Financial Economics* 9: 321–346.

Duffie D. (1989): *Futures Markets*, Prentice-Hall, Inc., Englewood Cliffs,
    New Jersey, 07632.

Engle R. (1982): "Autoregressive Conditional Heteroskedasticity with Es-
    timates of the Variance of United Kingdom Inflation", *Econometrica*
    50: 987–1007.

Fama E., and K. French (1987): "Commodity Futures Prices: Some Evidence on Forecast Power, Premiums and the Theory of Storage", *Journal of Business* 60: 55–74.

Fama E., and K. French (1988): "Business Cycles and the Behavior of Metals Prices", *Journal of Finance* 43(5): 1075–1094.

Jorion P. (1987): "On Jump Processes in the Foreign Exchange and in the Stock Market", unpublished paper, Columbia University, May.

Kaldor N. (1939): "Speculation and Economic Stability", *The Review of Economic Studies* 7: 1–27.

Miller M., and C. Upton (1985): "A Test of the Hotelling Valuation Principle", *Journal of Political Economy* 93(1): 1–25.

Ramaswamy K., and S. Sundaresan (1985): "The Valuation of Options on Futures Contracts", *Journal of Finance* 40: 1319–1340.

Schwartz E. (1982): "The Pricing of Commodity-Linked Bonds", *Journal of Finance* 37(2): 525–539.

Scott L. (1987): "Option Pricing When the Variance Changes Randomly: Theory, Estimation and An Application", *Journal of Financial and Quantitative Analysis* 22(4): 419–438.

Verleger P., jr. (1982): *Oil Markets in Turmoil: An Economic Analysis*, Ballinger Publishing Company, Cambridge, Massachusetts.

Wiggins J. (1987): "Stochastic Volatility Option Valuation: Theory and Empirical Estimates", *Journal of Financial Economics* 19(2): 351–372.

Working H. (1948): "The Theory of the Price of Storage", *American Economic Review* 39: 1254–1262.

Stochastic Models and Option Values
D. Lund and B. Øksendal (Editors)
© Elsevier Science Publishers B.V. (North-Holland), 1991

The Cost of a Promise to Develop an Oil Field within a Fixed Future Date

Petter Bjerksund

Institute of Finance and Management Science
Norwegian School of Economics and Business Administration
N-5035 Bergen, Norway

1. Introduction

In the spring of 1989, the Norwegian economy is characterized by depression, as contrasted by the world economy in general. The unemployment has risen dramatically to a post WW2 record. This situation is causing major concern among the citizens of Norway, leading to an increasing pressure on the political system to undertake concrete actions aimed at reducing unemployment.

Norway is in possession of large off-shore petroleum reserves. The activity in the petroleum sector is to a great extent influenced by political decisions. Not surprisingly, the idea to increase the activity in the petroleum sector by political means in order to reduce unemployment is being seriously discussed.

We analyze the case where the Government for some reason has promised to take steps so that the development of a particular oil reserve, currently held by a commercial company, will be initiated not later than a fixed future date. The purpose of the paper is to analyze some possible practical arrangements, and to evaluate the economic cost induced.

We focus the output price uncertainty as the risk dimension. The owner of the reserve may defer the irrevocable development decision. Development is modeled as undertaking a commitment to pay the future costs and to receive the future output according to the fixed preset production schedule. We interpret the Governmental promise as a constraint, as the set of possible strategies available to the present or to the future owner of the reserve is reduced. In this setting, the economic cost of the promise is identical to the opportunity loss caused by the constraint.

I thank Knut Kristian Aase, Steinar Ekern, Bruce Grundy, Diderik Lund, and Robert McDonald for helpful comments and discussions.

The paper is organized as follows. In section 2, our basic assumptions with respect to the economy and the project are stated, and some implications are discussed.

In section 3, we consider a "traditional" decision model. The owner of the field is in this case supposed to choose between accepting the project, rejecting it, or — once and for all — fixing a future date at which development will be initiated. We present the optimal decision rule and the field value, as well as the cost of the Governmental promise, according to this model.

In section 4 we implement the framework of contingent claims analysis, and take explicitly into account the opportunity to defer the development decision itself. First, we consider an oil reserve where the development decision in principle may be deferred perpetually. The oil reserve is interpreted as a perpetual American call option, and an explicit solution to the optimal strategy and the value is found by using known results.

Second, we analyze an arrangement where the Government declares to buy back the oil field at the market value and to undertake immediate development if the oil reserve is still undeveloped at the fixed future date. The promise then corresponds to undertaking a commitment to acquire a perpetual American call option at a fixed future date and exercising it immediately, conditional on the option still being "alive" at that date. The paper presents an explicit solution by which this commitment is evaluated.

Third, we consider an alternative arrangement where the Government is supposed to acquire the oil reserve immediately and to instruct the national oil company to start development not later than the fixed future date. In this case, the Governmental promise is taken into account when the company is managing the oil reserve.

Section 5 provides a numerical example. The results indicate that the traditional decision model substantially underestimates the opportunity loss from making the promise.

2. Assumptions

2.1 The economy

Our *first* basic assumption is that the dynamics of the spot price of oil is described by a geometric Brownian motion, defined by

$$\frac{dS(t)}{S(t)} = \alpha dt + \sigma dZ(t).$$ (1)

The term α represents a trend (if any), and σ is the instantaneous standard deviation of the relative price change per time unit. The term $dZ(t)$ corresponds to the increment of the standard Brownian motion.[1]

This assumption implies that the spot price of oil follows a continuous sample path, with no jumps. The spot price at a future date, as viewed from the current date, is log-normally distributed.[2]

Our *second* basic assumption is that there exists a tradable asset with relative return perfectly correlated with the relative return of the spot price of oil, and with identical volatility σ.[3] The relation between the equilibrium expected rate of return of the "twin asset", $\hat{\alpha}$, and the expected price change of the spot price of oil, α, is

$$\delta \equiv \hat{\alpha} - \alpha > 0 . \tag{2}$$

The term δ may be interpreted as the "rate of return shortfall" related to oil.[4] It represents the expected return forgone from holding oil if merely receiving return through the price change of oil. An analogy is δ being the continuous dividend pay-out rate on a stock.

Alternatively, δ may be considered as the "net marginal convenience yield" related to oil, reflecting the net benefit from the marginal unit of output in stock relative to a claim on future delivery of output.[5] With this interpretation, δ is similar to the liquidity premium from holding cash.

The assumption of $\delta > 0$ implies that there is an opportunity cost of having oil in the ground.

The risk free rate of interest $r > 0$ is constant and known. Riskless borrowing and lending at this rate r, are unrestricted.

Investors are assumed to prefer more to less, and to be risk averse. They are required to agree on the volatility σ of the spot price of oil. Investors

---

[1] For a brief introduction to stochastic calculus, see Smith (1979).

[2] The assumed price dynamics in eq. (1) may alternatively be stated

$$S(t) = S(0)e^{(\alpha - \frac{1}{2}\sigma^2)t + \sigma Z(t)} ,$$

where $Z(t)$ is the standard Brownian motion.

[3] The "twin asset" needs not necessarily to exist as a separate traded asset in the market. It is sufficient that its value can be replicated in the market by a portfolio of traded assets equipped with an appropriate dynamic self-financing strategy.

[4] See McDonald and Siegel (1984) and Pindyck (1988).

[5] Brennan and Schwartz (1985) use a constant rate δ when evaluating a copper mine.

need not necessarily to agree on the size of α or $\hat{\alpha}$, but only on their difference $\delta \equiv \hat{\alpha} - \alpha$.

We assume that the economy is frictionless, with continuous trading, and no sources to imperfections such as taxes, transaction costs, or short sale restrictions. The economy is characterized by no risk-free arbitrage opportunities.

To rule out riskless arbitrage opportunities, the prices in this economy must conform to the following evaluation rules: First, the value at date t of receiving one riskless dollar at the future date t' is

$$V_t[Y(t') = 1] = e^{-r(t'-t)}, \tag{3}$$

where $Y(t')$ is the certain future cash flow, and $V[\cdot]$ is a general evaluator.

Second, the value at date t of a claim on one unit of output at the future date t' is

$$V_t[Y(t') = S(t')] = e^{-\delta(t'-t)} S(t), \tag{4}$$

where $Y(t')$ is the random future cash flow, and $S(t)$ is the spot price at the evaluation date.[6]

2.2 The project

Our *third* basic assumption is that the project, once undertaken, may be described by a given production schedule, $q(\tau \mid t)$, and a given cost schedule, $b(\tau \mid t)$, where t is the initiation date and $\tau \geq 0$ is the project time. The project is irrevocable once undertaken.[7] Moreover, when development is initiated, there is no flexibility to reschedule the production or to abandon the project. The development decision may in principle be deferred perpetually.

We assume that if investments are initiated at some future date t' rather than at the current date t, the production schedule, considered as a function of project time τ, is unchanged. That is,

$$q(\tau \mid t) = q(\tau). \tag{5}$$

---

[6] In this economy, the equilibrium futures price at date t of a hypothetical futures contract with delivery date t' is

$$F_t(S(t), t') = e^{(r-\delta)(t'-t)} S(t),$$

see Brennan and Schwartz (1985) and Ross (1978).

[7] Majd and Pindyck (1987) consider the case where it takes time to build, and where the investment decision is a sequence of decisions rather than one irrevocable decision.

Furthermore, initiating at the future date t' in stead of the current date t causes the cost schedule to shift upwards at the exponential rate $\pi < r$. Stated formally, we assume

$$b(\tau \mid t') = e^{\pi(t'-t)}b(\tau \mid t),\tag{6}$$

where t' and t are alternative initiation dates.

Now, consider a project identical to the one described just above, where the irrevocable investment decision just has been made. In this case, there is by assumption no decision flexibility left. The oil field may thus be interpreted as a claim on future delivery of oil according to the fixed production schedule in eq. (5), combined with an obligation to repay a loan incurred according to the cost schedule in eq. (6).

By using the two evaluation rules, eqs. (3) and (4), it is easy to verify that at the current date t, the value of a commitment to initiate immediately is

$$C_t(t) = AS(t) - B(t),\tag{7}$$

where we define

$$A \equiv \int_0^\infty e^{-\delta\tau}q(\tau)d\tau,\tag{8}$$

and

$$B(t) \equiv \int_0^\infty e^{-r\tau}b(\tau \mid t)d\tau.\tag{9}$$

The constant A may be interpreted as the time-adjusted quantity of oil. It represents the quantity of oil received immediately that is equivalent to receiving the total quantity oil in the field, $Q \equiv \int q(\tau)d\tau$, according to the fixed production schedule $q(\tau)$. The constant $B(t)$ is the present value of future investment and production costs at date t, given immediate development.[8]

Note that the current value of the oil field is linear in the current spot price of oil $S(t)$. This is a consequence of no decision flexibility being left, and thus no opportunity for the manager to respond to oil price changes.

---

[8] The assumption of an exponentially escalating cost schedule, stated in eq. (6) above, implies

$$B(t') = e^{\pi(t'-t)}B(t),$$

where t' and t are alternative initiation dates.

2.3 Decision flexibility

Suppose that the owner is to choose between undertaking the project immediately, or rejecting it. The break-even price, $S_{BE}(t)$, corresponding to the current spot price at which immediate initiation has zero present value, is

$$S_{BE}(t) = \frac{B(t)}{A}, \tag{10}$$

where A and $B(t)$ are defined by eqs. (8) and (9) above, respectively. However, with the possibility of deferring the investment decision, and with (output price) uncertainty present, there is no reason to expect that the decision maker should restrict himself to choose between the two alternatives "accept" and "reject" only.

3. Traditional decision models

Consider the case where the owner of the field is facing the three following decision alternatives: To accept the project immediately, to fix — once and for all — a future date at which the project will be initiated, or to reject it. The decision flexibility thus corresponds to an immediate choice between investment strategies that are uniquely determined in calendar time.

The value of the oil field, contingent on the optimal decision being made, is given by

$$V(S,t) = \max\{C_t(t), \max_T C_t(T)\text{s.t. } T > t, 0\}. \tag{11}$$

The function $C_t(T)$ represents value of the oil field at the decision date t from undertaking a commitment to initiate development at the future date T.

By using footnote 8 and the evaluation rules in subsection 2.1, it is easy to verify that $C_t(T)$ is

$$C_t(T) \equiv V_t[C_T(T)] = e^{-\delta(T-t)}AS(t) - e^{-(r-\pi)(T-t)}B(t). \tag{12}$$

Depending on the parameter values, we either have the "optimal commitment timing" model or are left with the simple "accept/reject" model.

3.1 Optimal commitment timing

Assume for the moment $0 < \delta < r - \pi$. In the unconstrained case, the optimal decision is

$$\begin{array}{ll} \text{initiate at date } T^* & \text{if } S(t) < S^*_{OT}(t), \\ \text{accept} & \text{if } S(t) \geq S^*_{OT}(t). \end{array} \tag{13}$$

The optimal future initiation date[9] $T^* > t$ is defined by

$$T^* = \frac{\ln(S^*_{OT}(t)/S(t))}{(r - \pi) - \delta} + t, \tag{14}$$

and is a decreasing function of the current spot price $S(t)$. The critical price, indicating $T^* = t$, is

$$S^*_{OT}(t) = \frac{r - \pi}{\delta} S_{BE}(t), \tag{15}$$

where $S_{BE}(t)$ is the accept/reject break-even price. The value of the oil field according to the optimal timing model is thus

$$V(S(t), t) = \begin{cases} C_t(T^*) & \text{if } S(t) < S^*_{OT}(t), \\ C_t(t) & \text{if } S(t) \geq S^*_{OT}(t), \end{cases} \tag{16}$$

in the unconstrained case.

The constraint, induced by the Government's promise that the development of the oil field will be initiated within the future date T', is binding only if $T' < T^*$. The spot price, associated with $T^* = T'$, is

$$S(t) = f S^*_{OT}(t), \tag{17}$$

---

[9] The optimal future committed initiation date T^* is determined by the first-order condition

$$-\delta e^{-\delta(T^* - t)} A S(t) + (r - \pi) e^{-(r-\pi)(T^* - t)} B(t) = 0.$$

The solution $T^* > t$ (if any) is unique. With $0 < \delta < r - \pi$, it is easy to see that we have

$$\frac{\partial^2 C_t(T^*)}{\partial T^2} = (\delta - (r - \pi)) \delta e^{-\delta(T^* - t)} A S(t) < 0,$$

and $T^* > t$ (if any) thus represents a maximum.

where the factor f is

$$f \equiv e^{-[(r-\pi)-\delta](T'-t)} < 1 \,. \tag{18}$$

The value of the oil field according to the optimal timing model is thus

$$\overline{V}(S(t),t) = \begin{cases} C_t(T') & \text{if } S(t) < fS^*_{OT}(t), \\ C_t(T^*) & \text{if } fS^*_{OT}(t) \le S(t) < S^*_{OT}(t), \\ C_t(t) & \text{if } S^*_{OT}(t) \le S(t), \end{cases} \tag{19}$$

in the constrained case.

Our results are contained in the following table:

Parameter values: $0 < \delta < r - \pi$			
Spot price	Field value		Opportunity loss
S	V	$\overline{V}$	$V - \overline{V}$
$S < fS^*_{OT}$	$C_t(T^*)$	$C_t(T')$	$C_t(T^*) - C_t(T')$
$fS^*_{OT} \le S < S^*_{OT}$	$C_t(T^*)$	$C_t(T^*)$	0
$S^*_{OT} \le S$	$C_t(t)$	$C_t(t)$	0

We see that the optimal commitment timing model indicates an opportunity loss only when the current spot price $S(t)$ is lower than $fS^*_{OT}(t)$. If the spot price is higher, the model suggests that the promise is without cost.

3.2 Accept/reject

Assume instead $0 < r - \pi \le \delta$. Fixing a future date at which development is initiated will then never be the superior decision[10], and the problem collapses into the accept/reject model. In the unconstrained case, we then have that the optimal decision rule is

$$\begin{aligned} \text{reject} \quad & \text{if } S(t) < S_{BE}(t) \,, \\ \text{accept} \quad & \text{if } S(t) \ge S_{BE}(t) \,, \end{aligned} \tag{20}$$

---

[10] From eq. (12), we have that $\lim_{T \to \infty} C_t(T) = 0$, and the limit is thus equivalent to rejecting the project. If $r - \pi = \delta > 0$, eq. (12) boils down to $C_t(T) = e^{-\delta(T-t)}C_t(t)$, and we have the accept/reject situation. If $r - \pi < \delta$, the solution $T^* > t$ of the first-order condition (if any) represents a minimum, as $\partial^2 C_t(T^*)/\partial T^2 > 0$, and the accept/reject situation appears again.

where $S_{BE}(t)$ is the break-even price, defined by eq. (10). This leads to the following field value

$$V(S(t), t) = \begin{cases} 0 & \text{if } S(t) < S_{BE}(t), \\ C_t(t) & \text{if } S(t) \geq S_{BE}(t). \end{cases} \quad (21)$$

In the constrained case, rejecting the project is not feasible, as the development must be initiated not later than the future date T'. With the restrictions on the parameter values, fixing an initiation date T, with $t < T < T'$, is never the superior decision[11], and the choice is thus between accepting the project immediately, or initiating the project at the future date T'. The spot price, indicating indifference between accepting the project immediately and fixing the future initiation date T', is

$$S(t) = g S_{BE}(t), \quad (22)$$

where the factor g is

$$g \equiv \frac{1 - e^{-(r-\pi)(T'-t)}}{1 - e^{-\delta(T'-t)}} < 1. \quad (23)$$

If the current spot price of oil $S(t)$ is lower than $g S_{BE}(t)$, it is optimal to initiate the project at the future date T'. The value of the field in the constrained case is thus

$$\overline{V}(S(t), t) = \begin{cases} C_t(T') & \text{if } S < g S_{BE} \\ C_t(t) & \text{if } S \geq g S_{BE} \end{cases} \quad (24)$$

Our findings are contained in the following table:

Parameter values: $0 < r - \pi \leq \delta$			
Spot price	Field value		Opportunity loss
S	V	$\overline{V}$	$V - \overline{V}$
$S < g S_{BE}$	0	$C_t(T')$	$-C_t(T')$
$g S_{BE} \leq S < S_{BE}$	0	$C_t(t)$	$-C_t(t)$
$S_{BE} \leq S$	$C_t(t)$	$C_t(t)$	0

The accept/reject decision model suggests that the promise induces economic costs only if the current spot price $S(t)$ is below the break-even price $S_{BE}(t)$.

---

[11] From the unconstrained case, we know that the interior solution (if any) represents a minimum. The optimal value is thus found on one of the boundaries, and is given by $\max\{C_t(t), \lim_{T\to\infty} C_t(T)\}$. To find the optimal decision in the constrained case, it is thus sufficient to examine the values $C_t(t)$ and $C_t(T')$.

4. Contingent claims analysis

In the traditional decision model above, the owner is assumed to face an immediate choice between accepting the project, rejecting it, or fixing — once and for all — a future initiation date. With uncertainty present, however, deferring the *investment decision itself* is superior to fixing the initiation date in advance, as the future decision may be based on new information received in the meantime. To take this additional decision flexibility into account, we now implement the framework of contingent claims analysis.

4.1 The perpetual investment opportunity

Assume that the development decision in principle may be deferred perpetually. No cash outlays are incurred if the decision is deferred, but the present value of the future costs $B(t)$ escalates exponentially with rate π, see footnote 8.

We interpret the oil reserve as a contingent claim, with oil (spot) as the underlying asset.[12] The value of this perpetual investment opportunity, $U(S,t)$, satisfies the partial differential equation

$$\frac{1}{2}\sigma^2 S^2 U_{SS} + (r - \delta)SU_S - rU + U_t = 0\,,\tag{25}$$

where the subscripts indicate partial derivatives, see Merton (1977).

We next turn to the boundary conditions. First, at any date $\tau \in [t, \infty)$, there exists a critical spot price $S^*(\tau)$, at which the value of the investment opportunity is equal to the value of immediate development. That is

$$U(S^*(\tau)) = AS^*(\tau) - B(\tau)\,.\tag{26}$$

Furthermore, given the optimal investment strategy, the "high contact" or "smooth pasting" condition is also met, see Merton (1973), footnote [60]. The condition states that when the spot price equals the optimal critical price, the sensitivity of the asset value with respect to the spot price is independent of whether the investment opportunity is exercised or not. It this case the condition translates into

$$U_S(S^*) = A\,.\tag{27}$$

---

[12] Paddock, Siegel, and Smith (1988) consider an offshore petroleum lease as a contingent claim, using one unit of developed hydrocarbon reserve as the underlying asset. To obtain numerical results, the authors have to resort to a dubious "one-third" rule of thumb outside their model.

Finally, if the spot price approaches zero, this will also happen with the value of the investment opportunity. We thus have the third boundary condition

$$\lim_{S \to 0} U(S) = 0 \,. \tag{28}$$

Eqs. (25)–(28) determine the value of the investment opportunity $U(S,t)$ and the associated critical price $S^*(t)$. Our problem corresponds to a perpetual American call option written on a stock paying a continuous proportional dividend, see Samuelson (1965). McDonald and Siegel (1986) analyze a similar option with a stochastic exercise price.

The solution to our problem may be found by a suitable reinterpretation of parameters in one of the models of the two articles just mentioned, and is presented below. This solution satisfies the general partial differential equation of a contingent claim, see appendix A.

The critical price, indicating immediate investment, is

$$S^*(t) = \frac{\beta}{\beta - 1} S_{BE}(t) \,, \tag{29}$$

where the parameter $\beta > 1$ is defined by eq. (32) below. We see that the trigger price at date t, $S^*(t)$, corresponds to the accept/reject break-even price at that date, $S_{BE}(t)$, adjusted upwards with the factor $\beta/(\beta-1) > 1$.

The optimal decision is thus

$$\begin{aligned} \text{defer} \quad &\text{if } S(t) < S^*(t) \,, \\ \text{accept} \quad &\text{if } S(t) \geq S^*(t) \,. \end{aligned} \tag{30}$$

Note that the optimal strategy consists of a comparision between the current spot price of oil $S(t)$ and the trigger price $S^*(t)$.

The value of the investment opportunity, contingent on following the optimal investment strategy stated above, is

$$U(S(t), t) = \begin{cases} \alpha(t) S(t)^\beta & \text{if } S(t) < S^*(t), \\ AS(t) - B(t) & \text{if } S(t) \geq S^*(t), \end{cases} \tag{31}$$

where the parameter $\alpha(t) > 0$ is defined by eq. (33) below. As we would expect, $U(S,t)$ is positive, and increasing in the current spot price of oil, S. If the current spot price exceeds (or equals) the trigger price, immediate investment is optimal.

The exponent β and the constant α are defined by

$$\beta = \left(\frac{1}{2} - \frac{(r-\pi)-\delta}{\sigma^2}\right) + \sqrt{\left(\frac{(r-\pi)-\delta}{\sigma^2} - \frac{1}{2}\right)^2 + 2\frac{r-\pi}{\sigma^2}}, \qquad (32)$$

and

$$\alpha(t) = \frac{B(t)}{\beta - 1}S^*(t)^{-\beta}, \qquad (33)$$

respectively. The condition $\delta > 0$ ensures $\beta > 1$.

4.2 Possible future Governmental intervention

In this subsection, we introduce the Governmental promise that the development of the oil field, currently held by a commercial company through a perpetual licence, will be initiated within a future date T'. Assume that both parties have accepted that a possible transfer of the oil field is to be based on a mutual fair value. On one hand, the Government is required to refrain from passing on new legislation that "changes the rules of the game". To force the company to initiate development, or to expropriate the field, is thus not possible. On the other hand, the company is not entitled to reject an offer from the Government to buy back the undeveloped oil field at the market value.

Furthermore, assume that the Government declares at date t the strategy of buying back the oil field at the future date T' if the field is still undeveloped at that date, and in that case undertake immediate development. With this arrangement, the future status of the oil field determines whether future costs are incurred or not, while the spot price at date T' determines the level of the future costs (if any).

The net future cost of the transaction at date T', conditional on the field still being undeveloped at that date, is

$$X(T') = \alpha(T')S(T')^\beta - AS(T') + B(T'). \qquad (34)$$

The first term is the future market value of the perpetual investment opportunity. The two last terms represents the value of the field at date T' contingent on immediate initiation. We have defined $X(T')$ as a positive number.

The oil company will not suffer any loss from this possible future transaction, as the price paid by the Government is equal to the market price of the perpetual licence. During the period from date t to T', the commercial company will thus manage the investment opportunity as if it were a perpetual one. From subsection 4.1 above, we know that the optimal strategy is to initiate development immediately whenever the spot price of oil $S(\tau)$ hits the trigger price $S^*(\tau)$ from below.[13]

The status of the oil field at date T' is thus dependent on whether the spot price of oil $S(\tau)$ hits the trigger price $S^*(\tau)$ during the period $\tau \in [t, T']$ or not. Formally, the current cost of the Governmental promise, L, may be expressed as

$$L \equiv V_t\left[X(T') \cdot I\left(\inf_{\tau \in [t,T']} (S^*_\tau - S(\tau)) > 0\right)\right], \tag{35}$$

where $I(\cdot)$ is the indicator function, $V_t[\cdot]$ is a general evaluator, $X(T')$ is given by eq. (34), and where we for notational convenience denote the trigger price $S^*(\tau)$ by S^*_τ.

In this particular case, there exists an explicit solution to the current cost L. The solution may be written

$$L(S,t) = \alpha(T')\varphi(S,t \mid S^*_{T'}, \beta) - A\varphi(S,t \mid S^*_{T'}, 1) \\ + B(T')\varphi(S,t \mid S^*_{T'}, 0), \tag{36}$$

where the parameters $\alpha(T')$ and $B(T')$ are evaluated at the future date T'. By comparing eqs. (34) and (36), we see that the structure of $X(T')$ and $L(S,t)$ are similar.

The function φ represents the current value of a pay-off at the future date T according to a power-function of the spot price at that date, $S(T)^\varepsilon$, received if and only if $S(\tau)$ is lower than S^*_τ for all dates $\tau \in [t, T]$. The function φ is given by

$$\varphi(S,t \mid S^*_T, \varepsilon) \equiv V_t\left[S(T)^\varepsilon \cdot I\left(\inf_{\tau \in [t,T]} \left(e^{\pi(\tau-t)}S^*_t - S(\tau)\right) > 0\right)\right] \\ = \begin{cases} \Psi(S,t \mid S^*_T, \varepsilon) - (S^*_T)^{\varepsilon-\kappa}\Psi(S,t \mid S^*_T, \kappa) & \text{if } S(t) < S^*_t, \\ 0 & \text{if } S(t) \geq S^*_t, \end{cases} \tag{37}$$

where

$$\kappa(S,t \mid \varepsilon) = \varepsilon + 2\frac{\ln(S^*_t/S)}{\sigma^2(T-t)}. \tag{38}$$

---

[13] It follows from the definitions of $S^*(t)$ and $S_{BE}(t)$, eqs. (29) and (10), and footnote 8, that

$$S^*(t') = e^{\pi(t'-t)}S^*(t).$$

where t' and t are alternative initiation dates.

The function Ψ represents the current value of a pay-off at the future date T according to a power-function of the spot price at that date, $S(T)^\varepsilon$, received if and only if $S(T) < S_T^*$. The function Ψ is given by

$$\Psi(S, t \mid S_T^*, \varepsilon) \equiv V_t[S(T)^\varepsilon \cdot I(S(T) < S_T^*)]$$
$$= e^\lambda S^\varepsilon N[-d] \tag{39}$$

where

$$\lambda(t \mid \varepsilon) \equiv \left[(\varepsilon - 1)r - \varepsilon\delta + \frac{1}{2}\varepsilon(\varepsilon - 1)\sigma^2 \right](T - t), \tag{40}$$

$$d(S, t \mid S_T^*, \varepsilon) \equiv \frac{\ln(S/S_T^*) + \left[r - \delta + \left(\varepsilon - \frac{1}{2} \right)\sigma^2 \right](T - t)}{\sigma\sqrt{T - t}} \tag{41}$$

Consider the function Ψ above, and assume for the moment that the exponent $\varepsilon = 1$. In that case, Ψ represents the current value of receiving the risky underlying asset at the future date T, if and only if $S(T) < S_T^*$. With $\varepsilon = 0$, Ψ corresponds to the value of a one-dollar claim with maturity date T, conditional on the event $S(T) < S_T^*$.

We may use the evaluator Ψ to obtain the value of more complex future fixed-date pay-offs. Consider for instance a European put option written on S with exercise price S_T^* and maturity date T. The option represents the right, but no obligation, to sell one unit of the risky asset at the future date T at the price S_T^*. Given optimal exercise, the option is equivalent to a future obligation to deliver one unit of S at date T, if and only if $S(T) < S_T^*$, combined with a claim on S_T^* dollars with identical maturity and contingent on the same event. By using Ψ to evaluate the two simple contingent claims, we have that the current value of the European put option, P, is

$$P = S_T^* \Psi(S, t \mid S_T^*, 0) - \Psi(S, t \mid S_T^*, 1)$$
$$= e^{-r(T-t)} S_T^* N\left[-d(S, t \mid S_T^*, 0) \right]$$
$$- e^{-\delta(T-t)} S N\left[-d(S, t \mid S_T^*, 1) \right].$$

With $\delta = 0$, we have the Black-Scholes put option pricing formula, see, e.g., Ingersoll (1987) p. 320.

The function φ is somewhat more complex, as it gives the value of a future power-function pay-off $S(T)^\varepsilon$ received at date T, contingent on the spot price $S(\tau)$ being lower than the trigger price S_τ^* for *all dates* $\tau \in [t, T]$.

From eq. (37), we see that φ corresponds to the value of its counterpart only dependent on the event $S(T) < S_T^*$, Ψ, minus a "discount". This discount reflects that the claim considered here provides no pay-off at the future date T if the spot price $S(\tau)$ hits the trigger price S_τ^* in the meantime. With the evaluator φ at hand, we are provided with the necessary tool to find the current cost of the Governmental promise in this case.[14]

The cost of the Governmental promise, $L(S,t)$, satisfies the partial differential equation of a contingent claim. Furthermore, it can be shown that the cost of the promise has the following two limiting values:

$$\lim_{S\to 0} L(S,t) = e^{-r(T'-t)}B(T').$$

$$\lim_{t\to T'}\Big|_{S(t)<S_t^*} L(S,t) = \alpha(T')S(T')^\beta - AS(T') + B(T').$$

The first limit states that if the spot price approaches zero, the value of the promise approaches the current value of a commitment to pay the cost $B(T')$ at the future date T'. The second limit states that contingent on the spot price being lower than the corresponding trigger price, the value of the promise approaches the value of the immediate transaction as the "time to maturity" approaches zero.

4.3 Immediate Governmental intervention

The arrangement considered in the previous subsection may be viewed as somewhat "myopic", as the oil field is managed by the commercial company as if no promise exists during the period from date t to T'. Now, suppose instead that the Government buys the perpetual investment opportunity immediately at the current market value, and transfers it to the national oil company with an instruction to start development not later than the future date T'.

To the national oil company, the transferred oil field represents an obligation to initiate within the date T'. The national oil company, maximizing the value of its assets, will then choose an investment strategy during the period considered that explicitly takes into account the constraint imposed by the instruction.

The oil field, given the Governmental instruction, may be interpreted as a contingent claim. Its value, $W(S,t)$, satisfies the partial differential

---

[14] Both Ψ and φ satisfy the partial differential equation of a contingent claim, see appendix B. Appendix C indicates how the two functions are derived. For a more detailed discussion of the evaluators Ψ and φ, see Bjerksund (1989).

equation

$$\frac{1}{2}\sigma^2 S^2 W_{SS} + (r - \delta)SW_S - rW + W_t = 0 \,. \tag{42}$$

Furthermore, boundary conditions must be specified.

Suppose for the moment that the current spot price of oil is close to zero. In that case, the probability that the project will be initiated at the latest possible date approaches one. The obligation to initiate within the date T' will then converge towards an obligation to pay the cost at this future date, and we thus have

$$\lim_{S \to 0} W(S,t) = -e^{-r(T'-t)}B(T') \,. \tag{43}$$

If development has not been initiated before the future date T', immediate development is required. We have

$$W(S,T') = AS(T') - B(T') \,. \tag{44}$$

Finally, there exists a function $S_W^*(\tau)$, $\tau \in [t,T']$ describing the investment strategy that maximizes the value of the oil field. This condition is given by

$$W(S_W^*(\tau),\tau) = AS_W^*(\tau) - B(\tau) \,. \tag{45}$$

Unfortunately, there exists no closed form solution to the field value in this case. For a given set of parameter values, however, the value $W(S,t)$ may be approximated by numerical methods.[15]

The net cost of buying the perpetual investment opportunity immediately at the market price, and transferring the field to the national oil company with an instruction to initiate within the future date T', is thus

$$\text{Opportunity loss } = U - W \,, \tag{46}$$

where the first term, U, is given in subsection 4.1.

---

[15] For a survey of numerical methods, see Geske and Shastri (1985).

5. A numerical example

Assume the following parameter values:

PARAMETER VALUES			
rate of return shortfall	δ	0.04	per year
interest rate	r	0.02	per year
cost escalation rate	π	0.00	per year
volatility	σ^2	0.06	
time to maturity	$T' - t$	4	years
discounted quantity	A	100	mill. barrels
present value of costs	B	1000	mill. USD

It is easy to verify from eq. (10) that the accept/reject break-even price is

$$S_{BE} = 10 \text{ USD/barrel.}$$

The parameter values imply that the exponent, given by eq. (32), is $\beta = 2$ and the flexibility factor $\beta/(\beta - 1) = 2$. The trigger price, indicating immediate exercise of the perpetual investment opportunity, is thus

$$S^* = 20 \text{ USD/barrel,}$$

cf. eq. (29).

In this case, the traditional decision model in section 3 collapses into the accept/reject decision case. The critical price, indicating indifference between immediate development and fixing the future initiation date $T' = t + 4$, is

$$gS_{BE} = 5.2 \text{ USD/barrel,}$$

see eq. (22). In the case where the oil field is transferred to the national company with an instruction to initiate before date T', the trigger price is

$$S_W^* = 6.8 \text{ USD/barrel.}$$

In the following table, we present the field values for several levels of the spot price, given different degrees of decision flexibility.

FIELD VALUE (mill. USD)				
Spot price	Trad. model		Option model	
	V	$\overline{V}$	U	W
2.17	0.00	-738.07	11.79	-738.06
2.95	0.00	-671.97	21.72	-671.64
4.00	0.00	-582.26	40.00	-580.22
5.43	0.00	-457.12	73.68	-451.44
7.37	0.00	-263.19	135.72	-263.19
10.00	0.00	0.00	250.00	0.00
13.57	357.21	357.21	460.50	357.21
18.42	842.02	842.02	848.26	842.02
25.00	1500.00	1500.00	1500.00	1500.00

The second and the third column gives the value of the field according to the accept/reject model in the unconstrained case and in the constrained case, respectively. Note that V and $\overline{V}$ differ only when the spot price S is lower than the break-even price $S_{BE} = 10$ USD/barrel.

In the fourth column, we find the value of the investment opportunity, considered as a perpetual option. The last column represents the field value given the Governmental instruction. It should be no surprise that we have $U \geq W$. Furthermore, we note that the difference between $\overline{V}$ and W is negligible.

The opportunity losses are given in the following table:

OPPORTUNITY LOSS (mill. USD)			
Spot	Trad	Option-model	
price	$V - \overline{V}$	L	$U - W$
2.17	738.07	749.86	749.85
2.95	671.97	693.70	693.35
4.00	582.26	622.25	620.22
5.43	457.12	534.09	525.12
7.37	263.19	430.03	398.91
10.00	0.00	314.26	250.00
13.57	0.00	190.66	103.30
18.42	0.00	47.58	6.24
25.00	0.00	0.00	0.00

We see from the second column that according to the traditional model, economic costs are induced only if the current spot price is lower than the break-even price $S_{BE} = 10$ USD/barrel.

Furthermore, we clearly see that the traditional model understates the opportunity loss of making the development promise. From the previous table, we recall that the numerical values of $\overline{V}$ and W are approximately equal. The difference in the opportunity losses between column two and column four is thus mainly caused by the traditional model failing to evaluate properly the flexibility related to the unconstrained investment opportunity. The results indicate that if the Government bases the decision to make a promise as considered above on a traditional decision model, this may lead to a serious loss of value.

Appendix A: The perpetual investment opportunity

The value of the perpetual investment opportunity, $U(S, t)$, is given in subsection 4.1 above. It can be verified that the partial derivatives of this function are

$$U_S = \beta \frac{U}{S} , \tag{47}$$

$$U_{SS} = \beta(\beta - 1)\frac{U}{S^2} , \tag{48}$$

$$U_t = \pi U - \beta \pi U , \tag{49}$$

when $S(t) < S_t^*$.

By inserting the partial derivatives into the partial differential equation of a contingent claim, given by

$$\frac{1}{2}\sigma^2 S^2 U_{SS} + (r - \delta)SU_S - rU + U_t = 0 \,,$$

we obtain

$$\left[\frac{1}{2}\sigma^2 \beta^2 + (r - \pi - \delta - \frac{1}{2}\sigma^2)\beta - (r - \pi)\right]U = 0 \,. \qquad (50)$$

The trivial solution $U = 0$ is not of interest here. Inside the square brackets we find a square expression in the exponent β. It is easy to verify that

$$\beta = \left(\frac{1}{2} - \frac{(r - \pi) - \delta}{\sigma^2}\right) + \sqrt{\left(\frac{(r - \pi) - \delta}{\sigma^2} - \frac{1}{2}\right)^2 + 2\frac{r - \pi}{\sigma^2}} \,,$$

stated as eq. (32) above, is a solution to eq. (50). We may thus conclude that $U(S, t)$ satisfies the general partial differential equation of a contingent claim.

Appendix B: Possible future intervention — the PDE

In this appendix, we discuss the model of subsection 4.2 in relation to the general partial differential equation of a contingent claim. First, we present the partial derivatives of the evaluator

$$\Psi(S, t \mid S_T^*, \varepsilon) \equiv V_t[S(T)^\varepsilon \cdot I(S(T) < S_T^*)] \,, \qquad (51)$$

and conclude that Ψ satisfies the PDE. Next, we consider the evaluator

$$\varphi(S, t \mid S_T^*, \varepsilon) \equiv V_t\left[S(T)^\varepsilon \cdot I\left(\inf_{\tau \in [t, T]}\left(e^{\pi(\tau - t)}S_t^* - S(\tau)\right) > 0\right)\right], \qquad (52)$$

and its partial derivatives. With φ also satisfying the PDE, we finally conclude that this also is true for the cost of the Governmental promise, $L(S, t)$.

B.1: Ψ satisfies the PDE

The evaluator Ψ is a function of the current price of the underlying asset S, and time t. It may be verified that the partial derivatives of Ψ are

$$\Psi_S = (\varepsilon - F)\frac{\Psi}{S}\,, \tag{53}$$

$$\Psi_{SS} = \left\{\varepsilon(\varepsilon-1) - 2\varepsilon F + \frac{Fd}{\sigma\sqrt{T-t}} + F\right\}\frac{\Psi}{S^2}\,, \tag{54}$$

$$\Psi_t = \left\{-(\varepsilon-1)r + \varepsilon\delta - \frac{1}{2}\varepsilon(\varepsilon-1)\sigma^2 \right.$$
$$\left. + \left(r - \delta + (\varepsilon-\frac{1}{2})\sigma^2 - \frac{\sigma d}{\sqrt{T-t}}\right)F\right\}\Psi\,, \tag{55}$$

where we define

$$F(S,t \mid S_T^*,\varepsilon) \equiv \frac{n[d(S,t \mid S_T^*,\varepsilon)]}{\sigma\sqrt{T-t}N[-d(S,t \mid S_T^*,\varepsilon)]}\,. \tag{56}$$

Above, $n[\cdot]$ and $N[\cdot]$ denote the standard normal density and distribution functions, respectively, and d is defined by eq. (41).

The partial differential equation is

$$\frac{1}{2}\sigma^2 S^2 \Psi_{SS} + (r-\delta)S\Psi_S - r\Psi + \Psi_t = 0\,.$$

By inserting the partial derivatives into the left hand side above, it can be verified that the equation is satisfied.

B.2: φ satisfies the PDE

It can be shown that the partial derivatives of φ are

$$\varphi_S = \Psi_S(S,t \mid S_T^*,\varepsilon) - (S_T^*)^{\varepsilon-\kappa}\Psi_S(S,t \mid S_T^*,\kappa)$$
$$- \left(2F - 2\frac{d}{\sigma\sqrt{T-t}}\right)\frac{\varphi}{S}\,, \tag{57}$$

$$\varphi_{SS} = \Psi_{SS}(S,t \mid S_T^*,\varepsilon) - (S_T^*)^{\varepsilon-\kappa}\Psi_{SS}(S,t \mid S_T^*,\kappa)$$
$$- \frac{2}{\sigma^2}\left(2F - 2\frac{d}{\sigma\sqrt{T-t}}\right)\left((\kappa-\frac{1}{2})\sigma^2 - \frac{\sigma d}{\sqrt{T-t}}\right)\frac{\varphi}{S^2}\,, \tag{58}$$

$$\varphi_t = \Psi_t(S,t \mid S_T^*,\varepsilon) - (S_T^*)^{\varepsilon-\kappa}\Psi_t(S,t \mid S_T^*,\kappa)$$
$$- \left(2F - 2\frac{d}{\sigma\sqrt{T-t}}\right)\frac{\ln(S/S_T^*)}{T-t}\varphi\,, \tag{59}$$

when $S(t) < S_t^*$. The parameters d and F correspond to

$$d \equiv d(S, t \mid S_T^*, \kappa),$$
$$F \equiv F(S, t \mid S_T^*, \kappa),$$

cf. eqs. (41) and (56).

By inserting eqs. (57)–(59) into the partial differential equation

$$\frac{1}{2}\sigma^2 S^2 \varphi_{SS} + (r - \delta)S\varphi_S - r\varphi + \varphi_t = 0,$$

the terms on the left hand side cancel out.

B.3: L *satifies the PDE*

The cost of the Governmental promise is

$$L(S, t) = \alpha(T')\varphi(S, t \mid S_{T'}^*, \beta) - A\varphi(S, t \mid S_{T'}^*, 1) + B(T')\varphi(S, t \mid S_{T'}^*, 0),$$

see eq. (36). The parameters $\alpha(T')$, A, and $C(T')$ are independent of both S and t, and may here be considered as constants. The current cost of the Govermental promise thus corresponds to a linear combination of functions, that each satisfies the general partial differential equation. We may then conclude that $L(S, t)$ also satisfies the general partial differential equation of a contingent claim.

Appendix C: The derivation of Ψ and φ

This appendix indicates how the functions Ψ and φ are derived. For a detailed proof, see Bjerksund (1989).

C.1: The function Ψ

Consider the future pay-off

$$Y = S(T)^\varepsilon \cdot I(S(T) < S_T^*). \tag{60}$$

We recall from footnote 2 that the price process is

$$S(T) = S(0)e^{(\alpha - \frac{1}{2}\sigma^2)T + \sigma Z(T)}. \tag{61}$$

The model corresponds to a Black-Scholes economy, and the value of our contingent claim in eq. (60) is thus given by

$$\Psi \equiv V_0[S(T)^\varepsilon \cdot I(S(T) < S_T^*)]$$
$$= e^{-rT} E_0[S(T)^\varepsilon \cdot I(S(T) < S_T^*)\xi] , \qquad (62)$$

where

$$\xi = \exp\left\{ -\left(\frac{\hat{\alpha} - r}{\sigma}\right) Z(T) - \frac{1}{2}\left(\frac{\hat{\alpha} - r}{\sigma}\right)^2 T \right\} , \qquad (63)$$

see, e.g., Aase (1988) and Cox and Huang (1987). In eq. (62), ξ may be interpreted as the implicit risk-adjustment function.

The product of the function ξ and the probability density function of $Z(T)$ can be written as

$$\xi(Z(T)) \cdot n\left[\frac{Z(T)}{\sqrt{T}}\right] = n\left[\frac{1}{\sqrt{T}}\left(Z(T) + \left(\frac{\hat{\alpha} - r}{\sigma}\right) T\right)\right] , \qquad (64)$$

and may be interpreted as the risk-adjusted probability density. By writing eq. (62) as an integral, inserting eqs. (61) and (63), making the transformation[16]

$$Z^*(T) = Z(T) + \left(\frac{\hat{\alpha} - r}{\sigma}\right) T ,$$

some rearranging leads to an expectation of a log-normal truncated variable. It can be verified that eqs. (39)–(41) is the solution to the evaluation problem.

C.2: *The function* φ

Consider the future pay-off

$$Y = S(T)^\varepsilon I(S(T) < S_T^*) I(A_T) , \qquad (65)$$

where we define the event

$$A_T \equiv \left\{ \inf_{\tau \in [0,T]} (e^{\pi\tau} S_0^* - S(\tau)) > 0 \right\} . \qquad (66)$$

---

[16] The transformation is related to the Girsanov theorem.

The pay-off at the future date T is made conditional on the price path of S during the period $[0, T]$. This contingent claim may be attained in our economy, see Harrison and Kreps (1979).[17]

The unique price at which the claim with the future pay-off in eq. (65) may be attained today, is

$$\varphi \equiv V_0[S(T)^\varepsilon I(S(T) < S_T^*)I(A_T)]$$
$$= e^{-rT} E_0[S(T)^\varepsilon I(S(T) < S_T^*)I(A_T)\xi] \,, \qquad (67)$$

where ξ is the implicit risk-adjustment function.

By using conditional expectation, we may alternatively write eq. (67) as

$$\varphi = e^{-rT} E_0[S(T)^\varepsilon I(S(T) < S_T^*) E_0[I(A_T)\xi \mid S(T)]] \,. \qquad (68)$$

We see from eq. (68) that given the terminal price $S(T)$, the sample paths for which $I(A_T) = 0$ do not contribute to the value of the claim. Given the terminal price $S(T)$ and $I(A_T) = 1$, the function ξ is identical to eq. (63). Note that ξ is determined by $S(T)$. The inner expectation of eq. (68) thus boils down to

$$E_0[I(A_T)\xi \mid S(T)] = \Pr\{A_T \mid S(T)\}\xi \,. \qquad (69)$$

It can be verified that the conditional probability is[18]

$$\Pr\{A_T \mid S(T)\} = 1 - \left(\frac{S_T^*}{S(T)}\right)^{\varepsilon - \kappa} \,, \qquad (70)$$

---

[17] The authors consider a general model where the economy is generated by a multi-dimensional diffusion process. They note (p. 396) that "every contingent claim may be expressed as a function of the vector price history over the interval $[0, t]$".

[18] The event A_T may equivalently be expressed

$$A_T = \left\{\inf_{\tau \in [0,T]} W(\tau) > 0\right\} \,,$$

where
$$W(\tau) \equiv \ln\left(S_\tau^*/S(\tau)\right)$$
$$= \ln\left(S_0^*/S(0)\right) + \left(\pi - \alpha + \frac{1}{2}\sigma^2\right)\tau - \sigma Z(\tau) \,,$$

and $Z(\tau)$ is the standard Brownian motion generating the price $S(\tau)$.

The conditional probability $\Pr\{A_T \mid S(T)\}$ thus corresponds to the probability of a Brownian motion (with drift), $W(\tau)$, being positive during the period $[0, T]$, given both the initial value $W(0)$ and the terminal value $W(T)$.

where we define

$$\kappa \equiv \varepsilon + 2\frac{\ln\left(S_0^*/S(0)\right)}{\sigma^2 T}\,. \tag{71}$$

By inserting eqs. (69) and (70) into eq. (68), and rearranging, we obtain

$$\varphi = e^{-rT}E_0[I(S(T) < S_T^*)S(T)^\varepsilon \xi]$$
$$- (S_T^*)^{\varepsilon-\kappa}e^{-rT}E_0[I(S(T) < S_T^*)S(T)^\kappa \xi]\,.$$

Finally, by using the definition of Ψ, see eq. (62), we have the result

$$\varphi(S,t \mid S_T^*,\varepsilon) = \Psi(S,t \mid S_T^*,\varepsilon) - (S_T)^{\varepsilon-\kappa}\Psi(S,t \mid S_T^*,\kappa)\,, \tag{72}$$

where κ is defined by eq. (71).

References

Aase K.K. (1988): "Contingent Claims Valuation when the Security Price is a Combination of an Ito Process and a Random Point Process", *Stochastic Processes and Their Applications* 28: 185–220.

Bjerksund P. (1989): "A Contingent Claims Analysis of an Oil Reserve", doctoral dissertation, Norwegian School of Economics and Business Administration.

Brennan M.J., and E.S. Schwartz (1985): "Evaluating Natural Resource Investments", *Journal of Business* 58: 135–157.

Cox J.C., and C. Huang (1987): "Option Pricing Theory and Its Applications", in S. Bhattacharya and G.M. Constantinides (editors), *Frontiers of Financial Theory*, Rowman & Littlefield.

Geske R., and K. Shastri (1985): "Valuation by Approximation: A Comparision of Alternative Option Valuation Techniques", *Journal of Financial and Quantitative Analysis* 20: 45–71.

Harrison M.J., and D.M. Kreps (1979): "Martingales and Arbitrage in Multiperiod Securities Markets", *Journal of Economic Theory* 20: 381–408.

Ingersoll J.E., jr. (1987): *Theory of Financial Decision Making*, Rowman & Littlefield.

McDonald R., and D. Siegel (1984): "Option Pricing When the Underlying Asset Earns a Below-Equilibrium Rate of Return: A Note", *Journal of Finance* 39: 261–265.

McDonald R., and D. Siegel (1986): "The Value of Waiting to Invest", *Quarterly Journal of Economics* 101: 707–727.

Majd S., and R.S. Pindyck (1987): "Time to Build, Option Value, and Investment Decisions", *Journal of Financial Economics* 18: 7–27.

Merton R.C. (1973): "Theory of Rational Option Pricing", *Bell Journal of Economics and Management Science* 4: 141–183.

Merton R.C. (1977): "On the Pricing of Contingent Claims and the Modigliani-Miller Theorem", *Journal of Financial Economics* 5: 241–249.

Paddock J.L., D.R. Siegel, and J.L. Smith (1988): "Option Valuation of Claims on Real Assets: The Case of Offshore Petroleum Leases", *Quarterly Journal of Economics* 103: 479–508.

Pindyck R.S. (1988): "Irreversible Investment, Capacity Choice, and the Value of the Firm", *American Economic Review* 78(5): 969–985.

Ross S.A. (1978): "A Simple Approach to the Valuation of Risky Streams", *Journal of Business* 51: 453–475.

Samuelson P.A. (1965): "Rational Theory of Warrant Pricing", *Industrial Management Review*, 6: 13–39.

Smith C.W., jr. (1979): "Applications of Option Pricing Analysis", in J.L. Bicksler (editor), *Handbook of Financial Economics*, North-Holland, 79–121.

Stochastic Models and Option Values
D. Lund and B. Øksendal (Editors)

Irreversibility and the Explanation of Investment Behavior

Robert S. Pindyck

Massachusetts Institute of Technology
Cambridge, MA 02139

1. Introduction

The explanation of aggregate and sectoral investment behavior has been one of the less successful endeavors in empirical economics. Existing econometric models have simply not been very good at explaining or predicting investment spending. The problem is not just that these models have been unable to explain and predict more than a small portion of the movements in investment. In addition, constructed quantities that in theory should have strong explanatory power — e.g. Tobin's q, or various measures of the cost of capital — in practice do not, and leave much of investment spending unexplained.[1]

It is easy to think of reasons for the failings of these models. For example, even leaving aside problems with their theoretical underpinnings, there are likely to be formidable estimation problems resulting from aggregation (across firms, and also across investment projects of different gestations). I will not attempt to survey these problems here, nor in any way provide a general overview of the state of investment modelling. Instead I want to focus on one special aspect of investment — the role of risk, and in particular the effects that risk has on investment when expenditures are irreversible.

Most major investment expenditures have two important characteristics which together can dramatically affect the decision to invest. First, the expenditures are largely irreversible; the firm cannot disinvest, so the expenditures are sunk costs. Second, most major investments can be

My thanks to Diderik Lund and Robert McDonald for their helpful comments. Financial support for this research was provided by M.I.T.'s Center for Energy Policy Research, and by the National Science Foundation under Grant No. SES-8318990.

[1] See Kopcke (1985) for an overview, as well as examples and comparisons of traditional approaches to modelling investment spending.

delayed. Hence the firm can wait for new information to arrive about prices, costs, and other market conditions before it commits resources.

Irreversibility usually arises because capital is industry or firm specific, i.e., it cannot be used productively in a different industry or by a different firm. A steel plant, for example, is industry specific. It can only be used to produce steel, so if the demand for steel falls, the market value of the plant will fall. Although the plant could be sold to another steel company, there is likely to be little gain from doing so, so the investment in the plant must be viewed as a sunk cost. As another example, most investments in marketing and advertising are firm specific, and so are likewise sunk costs. Partial irreversibility can also result from the "lemons" problem.[2] Office equipment, cars, trucks, and computers are not industry specific, but have resale value well below their purchase cost, even if new. In this case, at least a portion of the initial investment is a sunk cost.

When investments are irreversible and can be delayed, the decision to invest becomes extremely sensitive to uncertainty over future cash flows. In the next section I summarize the reasons why this is the case. This summary is quite brief, and largely refers to the literature on "real options" that has emerged on this topic during the past few years. A much more detailed summary and explanation of the analysis of irreversible investment can be found in my recent survey paper (1990). Here my focus is on the empirical modelling of investment (and the need for better empirical studies). Do risk and irreversibility interact in affecting investment behavior the way theory says they should? Sections 3 and 4 discuss empirical issues, and summarize two studies (one of which is mine) that represent the small amount of evidence on this question that I am aware of. Section 5 briefly discusses future research and concludes.

2. Implications of irreversibility

Non-diversifiable risk plays a role in even the simplest models of investment, by affecting the cost of capital. But there is an emerging literature that suggests that risk may be a more crucial explanator of investment. The thrust of this literature begins with the fact that much of investment spending is irreversible and can be delayed. In this case, the standard

---

[2] As Akerlof (1970) first demonstrated, when sellers of a good know more about its quality than do buyers, lowquality goods tend to drive highquality ones from the market. As a result, used goods will sell for less than they would in a world with perfectly informed buyers and sellers.

net present value method that evaluates an investment by comparing its cost to the present value of the cash flows it is expected to generate does not apply.

Instead, the investment expenditure can be viewed as analogous to the exercising of an option (an option to productively invest). Once such an option is exercised, it is "dead", i.e. one cannot decide to exercise it instead at some point in the future, or never at all. Specifically, one gives up the option of waiting for new information (about evolving demand and cost conditions), and using that information to reevaluate the desirability and/or timing of the expenditure.[3] This lost option value must be included as part of the cost of the investment. Doing so leads to an investment rule that can be viewed as a *modified* net present value rule: "Invest when the value of a project is at least as large as the direct cost of the project plus the opportunity cost of exercising the option to invest". In other words, the value of the project must exceed the cost by an amount equal to the value of keeping the firm's option to invest alive.[4]

Recent studies (see the references in Footnote 3) have shown that this opportunity cost can be large, and investment rules that ignore it can be grossly in error. Also, this opportunity cost is highly sensitive to uncertainty over the future value of the project, so that changing economic conditions that affect the perceived riskiness of future cash flows can have a large impact on investment spending, larger than, say, a change in interest rates. (In the context of macroeconomic policy, this means that if the goal is to stimulate investment, stability and credibility may be much more important than tax incentives or interest rates.)

---

[3] This is developed in the articles by Bernanke (1983) and McDonald and Siegel (1986). Other examples of this literature include Cukierman (1980), Brennan and Schwartz (1985), Majd and Pindyck (1987), Bertola (1989), Pindyck (1988), and Dixit (1989). In the articles by Bernanke and Cukierman, uncertainty over future market conditions is reduced as time passes, so that firms have an incentive to delay investing when markets are volatile. In the other papers, future market conditions are *always* uncertain. As with a call option on a dividend-paying stock, an investment expenditure should be made only when the value of the resulting project exceeds its cost by a positive amount, and again, increased uncertainty will increase the incentive to delay the investment.

[4] Firms obtain their options to invest in various ways. Sometimes they result from patents, or ownership of land or natural resources. More generally, they arise from a firm's managerial resources, technological knowledge, reputation, market position, and possibly scale, all of which may have been built up over time, and which enable the firm to productively undertake investments that individuals or other firms cannot undertake. Most important, these options to invest are valuable. Indeed, for most firms, a substantial part of their market value is attributable to their options to invest and grow in the future, as opposed to the capital that they already have in place. See Kester (1984) and Pindyck (1988).

The dependence of investment on interest rates provides an interesting example of the implications of irreversiblity. Interest rates are key variables in traditional econometric models of investment, but they are often found to be statistically insignificant when those models are estimated. In a recent paper, Ingersoll and Ross (1988) examined irreversible investment decisions when the interest rate evolves stochastically, but future cash flows are known with certainty. As with uncertainty over future cash flows, this creates an opportunity cost of investing, so that the traditional NPV rule will accept too many projects. Instead, an investment should be made only when the interest rate is below a critical rate, r^*, which is *lower* than the internal rate of return, r^0, which makes the NPV zero. Also, the difference between r^* and r^0 grows as the volatility of interest rates grows.

Ingersoll and Ross also show that for longlived projects, a decrease in expected interest rates for all future periods need not accelerate investment. The reason is that such a change also lowers the cost of waiting, and thus can have an ambiguous effect on investment. This shows how the level of interest rates may be of only secondary importance as a determinant of aggregate investment spending. Interest rate volatility (as well as the volatility of other variables) may be more important.

As another example, consider the recessions of 1975 and 1980. The sharp jumps in energy prices that occurred in 1974 and 1979–80 clearly contributed to those recessions. They caused a reduction in the real national incomes of oil importing countries, and they led to "adjustment effects" — inflation and a further drop in real income and output resulting from the rigidities that prevented wages and non-energy prices from coming into equilibrium quickly. But those energy shocks also caused greater uncertainty over future economic conditions. In particular, it was unclear whether energy prices would fall or continue to rise, what the impact of higher energy prices would be on the marginal products of various types of capital, how long-lived the inflationary impact of the shocks would be, etc. Other events also made the economic environment more uncertain, especially in 1979–82 in the United States: much more volatile exchange rates and interest rates. This may have also contributed to the decline in investment spending that occurred during these periods.[5]

---

[5] This point was made by Bernanke (1983), particularly with respect to changes in oil prices. Also, see Evans (1984) and Tatom (1984) for a discussion of the depressive effects of increased interest rate volatility.

3. Neoclassical models of investment

As I suggested at the outset of this paper, the irreversibility of most investment expenditures, and the implications that irreversibility has for risk, may help to explain why neoclassical investment theory has failed to provide good empirical models of investment behavior. Effects of risk are typically handled by assuming that a risk premium (obtained, say, from the CAPM) can be added to the discount rate used to calculate the present value of a project. But as we have learned from financial option pricing and its application to real investment, the correct discount rate cannot be obtained without actually solving the option valuation problem, that discount rate need not be constant over time, and it will not equal the firm's average cost of capital. As a result, simple cost of capital measures, based on rates of return (simple or adjusted) to equity and debt, may be poor explanators of investment spending.

This can be seen in the context of models based on Tobin's q. A good example is the model of Abel and Blanchard (1986), which is one of the most sophisticated attempts to explain investment in a q theory framework; it uses a carefully constructed measure for marginal rather than average q, incorporates delivery lags and costs of adjustment, and explicitly models expectations of future values of explanatory variables.

The model is based on the standard discounted cash flow rule, "invest in the marginal unit of capital if the present discounted value of the expected flow of profits resulting from the unit is at least equal to the cost of the unit". Let $\pi_t(K_t, I_t)$ be the maximum value of profits at time t, given the capital stock K_t and investment level I_t, i.e. it is the value of profits assuming that variable factors are used optimally. It depends on I_t because of costs of adjustment; $\partial \pi/\partial I < 0$, and $\partial^2 \pi/\partial I^2 < 0$, i.e. the more rapidly new capital is purchased and installed, the more costly it is. Then the present value of current and future profits is given by:

$$V_t = E_t \left[\sum_{j=0}^{\infty} \left[\prod_{i=0}^{j} (1 + R_{t+i})^{-1} \right] \pi_{t+j}(K_{t+j}, I_{t+j}) \right], \qquad (1)$$

where E_t denotes an expectation, and R is the discount rate. Maximizing this with respect to I_t, subject to the condition $K_t = (1-\delta)K_{t-1} + I_t$ (where δ is the rate of depreciation), gives the following marginal condition:

$$-E_t \left(\frac{\partial \pi_t}{\partial I_t} \right) = q_t, \qquad (2)$$

where

$$q_t = E_t \left[\sum_{j=0}^{\infty} \left[\prod_{i=0}^{j} (1 + R_{t+i})^{-1} \right] \left(\frac{\partial \pi_{t+j}}{\partial K_{t+j}} \right) (1 - \delta)^j \right]. \qquad (3)$$

In other words investment occurs up to the point where the cost of an additional unit of capital equals the present value of the expected flow of incremental profits resulting from the unit. Abel and Blanchard estimate both linear and quadratic approximations to q_t, and use vector autoregressive representations of R_t and $\partial \pi_t / \partial K_t$ to model expectations of future values. Their representation of R_t is based on a weighted average of the rates of return on equity and debt.

If the correct discount rates R_{t+i} were known, eqns. (2) and (3) would indeed accurately represent the optimal investment decision of the firm. The problem is that these discount rates are usually not known, and generally will not equal the average cost of capital of the firm, or some related variable. Instead, these discount rates can only be determined as part of the solution to the firm's optimal investment problem. This involves valuing the firm's options to make (irreversible) marginal investments (now or in the future), and determining the conditions for the optimal exercise of those options. Thus the solution to the investment problem is more complicated than the first order condition given by (2) and (3) would suggest.

As an example, consider a project that has zero systematic (nondiversifiable) risk. The use of a risk-free interest rate for R would lead to much too large a value for q_t, and might suggest that an investment expenditure should be made, whereas in fact it should be delayed. Furthermore, there is no simple way to adjust R properly. The problem is that the calculation ignores the opportunity cost of exercising the option to invest.[6] This may be why Abel and Blanchard conclude that "our data are not sympathetic to the basic restrictions imposed by the q theory, even extended to allow for simple delivery lags".

4. Does risk matter

Unfortunately, incorporating irreversibility into models of aggregate investment spending is not a simple matter. First, the equations describing

---

[6] In principle, one could extend the Abel-Blanchard model to account for irreversibility. For example, one could introduce asymmetric adjustment costs, as Bizer and Sichel (1988) have done. (This is discussed in the next section.) However, in practice it would then be difficult to solve for marginal q and the optimal investment rule.

optimal investment decisions are extremely nonlinear, even for very simple models. (See my survey paper (1990), and the references in Footnote 3.) Second, it is difficult to measure (and sometimes even identify) the variables or parameters that reflect key components of risk. Partly as a result of this, there has been very limited empirical work to date that tests the importance of irreversibility for the modelling of investment spending. Here, I briefly survey the little empirical work that I am aware of.

One paper that provides a test of irreversibility and its implications is by Bizer and Sichel (1988). They develop a model of capital accumulation and utilization with asymmetric costs of adjustment, i.e., the costs of adjusting the capital stock up or down can differ. If irreversibility is important, one would expect to find that downward adjustment costs exceed upward ones. Bizer and Sichel derive an Euler equation, which they estimate using Hansen and Singleton's (1982) generalized instrumental variable procedure. They do not have firm data, so instead they use 2-digit SIC industry data. They measure asymmetry of adjustment costs with respect to two reference points: a zero level of investment, and a "normal" (average) level of investment.

Their preliminary results indicate some evidence of irreversibility, in particular in primary metals, fabricated metal products, and possibly the paper industry. But they also find that upward adjustment costs exceed downward ones in the food and petroleum industries. This may simply mean that aggregation is masking irreversibility. Other problems include the use of a single discount factor (the S&P dividend/price ratio, which is very volatile) and cost of capital for all industries. Nonetheless, their approach seems like a promising way to test for effects of irreversibility, particularly if used with disaggregated data. It does not, however, explicitly deal with effects of changes in risk.

In a recent working paper (1986), I performed some very simple nonstructural tests for the importance of risk. I used data on the stock market, on the grounds that when product markets become more volatile, we would expect stock prices to also become more volatile, so that the variance of stock returns will be larger. This was indeed the case, for example, during the recessions of 1975 and 1980, and most dramatically during the Great Depression. Thus the variance of aggregate stock returns should be correlated with aggregate product market uncertainty.

Stock returns themselves are also a predictor of aggregate investment spending. My concern was whether the *variance* of stock returns also has predictive power with respect to investment, and whether that predictive power goes beyond that of stock returns themselves, as well as other variables that would usually appear in an empirical investment equation. I conducted two related exploratory tests. First, I tested and was able to accept the hypothesis that the variance of stock returns Granger-causes the real growth rate of investment. Specifically, the variance of returns is a strong predictor of investment growth, but investment growth does not predict the variance of returns.[7] This is true when total fixed investment is used, and also when structures and equipment are treated separately.

Second, I ran a set of regressions similar to those used by Fischer and Merton (1984) in their study of the predictive power of stock returns. Each investment variable (the growth rates of total fixed investment, investmentin structures, and investment in equipment) is regressed first against lagged values of variance, then against lagged values of variance and lagged values of real stock returns, and finally against lagged variance, lagged stock returns, and the lagged values of four additional variables that often appear in empirical investment equations: the change in the BAA corporate bond rate, the change in the 3-month Treasury bill rate, the change in the rate inflation, and the rate of growth of real GNP.

The results are shown in Table 1. Note that three lagged values of each independent variable appear in the regressions, but the table only shows the sum of the estimated coefficients for each variable, together with an associated t-statistic. In all of the regressions, the quarterly variance of stock returns is computed using CRISP data on daily returns. Specifically, the variance for quarter t is computed as:

$$\sigma_t^2 = \frac{1}{n} \sum_{j=1}^{n} x_{t,j}^2 , \qquad (4)$$

where n is the number of days in the quarter, and $x_{t,j}^2$ is the logarithmic return from day $j - 1$ to day j in quarter t, adjusted for nontrading days by dividing by the square root of the number of days between trades.

Consistent with the causality tests mentioned above, the variance of returns is highly significant when it appears as the only independent variable. Variance continues to be highly significant after adding real

---

[7] To say that "X causes Y", two conditions should be met. First, X should help to predict Y, i.e. in a regression of Y against past values of Y, the addition of past values of X as independent variables should contribute significantly to the explanatory power of the regression. Second, Y should *not* help to predict X. (If X helps to predict Y *and* Y helps to predict X, it is likely that one or more other variables are in fact "causing" both X and Y.)

stock returns to the regression. This second independent variable is also a significant explanator of the growth of total non-residential investment, as Fischer and Merton found, as well as investment in equipment, but it is not significant for investment in structures. When the remaining explanatory variables are added, variance continues to be significant in the regressions for non-residential investment and structures, but not forequipment. (The only significant explanator of equipment investment is the BAA bond rate.) This may reflect the fact that the irreversibility of investment is greater for structures than for equipment.

The importance of this risk measure is also evident from the magnitudes of the variance coefficients. As I have shown elsewhere (1984), the quarterly variance of stock returns went from about .01 in the 1960's to about .02 in the mid-1970's. From regressions 3, 6, and 9 we see that this implies an approximately 4.5 percentage point decline in the growth rate of investment in structures (a drop from around 5 percent real growth during the 1960's to only slightly positive real growth), a 2.5 percentage point drop in the growth rate of investment in equipment, and a 3 percentage point drop in the growth rate of total investment.

Of course regressions of this sort are extremely crude, and are based on aggregate data and what is probably a very imperfect measure of risk. (Even if the variance of stock returns is a good proxy for the volatility of cash flows, it does not capture the "peso problem", i.e., changing perceptions of the risk associated with one or more possible future catastrophic events.) Nonetheless, the results suggest that the explicit inclusion of market risk measures may help to improve our ability to explain and predict investment spending, and that the development of structural models that include such measures should be an important research priority.

5. Conclusions

There are good theoretical reasons to expect market risk to have a major role in the determination of investment spending. This idea is not new; it has been elaborated upon in a number of articles during the past few years. However, it seems to be missing from most empirical work on investment. This may be a reflection of the fact that most theoretical models of irreversible investment under uncertainty are quite complicated, so that their translation into well-specified empirical models represents a formidable task.

Table 1: Variance of stock returns as a predictor of investment. (Quarterly data, 1963-4 to 1983-4.)

Indep.					Dependent Variable				
Var.	INGR	INGR	INGR	ISGR	ISGR	ISGR	IEGR	IEGR	IEGR
Reg. No.	1	2	3	4	5	6	7	8	9
CONST	.0314	.0264	.0199	.0261	.0228	.0190	.0349	.0291	.0201
	(7.72)	(6.30)	(2.38)	(5.36)	(4.28)	(1.75)	(6.96)	(5.67)	(2.01)
$\sum_{i=1}^{3}$ VAR$_{-i}$	-6.307	-5.306	-3.282	-6.262	-5.470	-4.786	-6.478	-5.378	-2.563
	(-5.98)	(-4.92)	(-2.32)	(-4.98)	(-4.20)	(-2.59)	(-4.99)	(-4.08)	(-1.51)
$\sum_{i=1}^{3}$ RTRN$_{-i}$		.1738	.1229		.0769	.0887		.2150	.1388
		(3.46)	(1.91)		(1.22)	(1.05)		(3.49)	(1.80)
$\sum_{i=1}^{3}$ DRBAA$_{-i}$			-.0308			-.0056			-.0409
			(-2.93)			(-0.41)			(-3.25)
$\sum_{i=1}^{3}$ DRTB3$_{-i}$			.0167			.0199			.0144
			(2.44)			(2.23)			(1.76)
$\sum_{i=1}^{3}$ DINF$_{-i}$			-.2827			-.9405			.1495
			(-0.41)			(-1.05)			(0.18)

Table 1: (Continued) Variance of stock returns as a predictor of investment. (Quarterly data, 1963–4 to 1983–4.)

Indep. Var.	Dependent Variable								
	INGR	INGR	INGR	ISGR	ISGR	ISGR	IEGR	IEGR	IEGR
Reg. No.	1	2	3	4	5	6	7	8	9
$\sum_{i=1}^{3}$ GNPGR$_{-i}$			.3505			.0816			.5428
			(0.66)			(0.18)			(0.86)
RHO	.0222	.0187	.0108	.0226	.0202	.0126	.0215	.0174	.0096
	(2.95)	(2.59)	(1.46)	(2.51)	(2.20)	(1.32)	(2.33)	(1.97)	(1.11)
R^2	.425	.501	.627	.320	.335	.477	.350	.442	.596
SER	.0204	.0194	.0184	.0244	.0247	.0239	.0252	.0238	.0216
DW	1.40	1.53	1.84	1.65	1.67	1.88	1.60	1.78	2.19

Variables: INGR = Quarterly growth rate of real business fixed investment. ISGR = Growth rate of real investment in structures. IEGR = Growth rate of real investment in durable equipment. VAR = Quarterly variance of real return on NYSE Index. RTRN = Real return on NYSE Index. DRBAA = Change in BAA corporate bond rate. DRTB3 = Change in 3-month Treasury bill rate. DINF = Change in inflation rate, as measured by PPI. GNPGR = Quarterly growth rate of real GNP.

Note: *t*-statistics in parentheses.

In any case, the gap here between theory and empiricism is somewhat disturbing. While it is clear from the theory that increases in the volatility of, say, interest rates or exchange rates should depress investment, it is not at all clear *how large* the effect is likely to be. Nor is it clear how important these factors have been as explanators of investment across countries and over time.

Determining the importance of these factors should be a research priority. One approach is to do empirical testing of the sort discussed in the preceding section, perhaps using cross section data for a number of countries. Another approach is to construct simulation models based on theoretical formulations that can be solved either analytically or numerically, and then parameterize them so that they "fit" particular industries. One could then calculate predicted effects of observed changes in, say, price volatility, and compare them to the predicted effects of changes in interest rates or tax rate. Simulation models of this sort could likewise be constructed to predict the effect of a perceived possible shift in the tax regime, the imposition of price controls, etc. Such models may also be a good way to study uncertainty of the "peso problem" sort.

References

Abel A.B., and O.J. Blanchard (1986): "The Present Value of Profits and Cyclical Movements in Investment", *Econometrica* 54: 249–273.

Akerlof G.A. (1970): "The Market for 'Lemons': Quality Uncertainty and the Market Mechanism", *Quarterly Journal of Economics* 84: 488–500.

Baldwin C.Y. (1982): "Optimal Sequential Investment when Capital is Not Readily Reversible", *Journal of Finance* 37: 763–782.

Bernanke B.S. (1983): "Irreversibility, Uncertainty, and Cyclical Investment", *Quarterly Journal of Economics* 98: 85–106.

Bertola G. (1989): "Irreversible Investment", unpublished working paper, Princeton University.

Bizer D.S., and D.E. Sichel (1988): "Irreversible Investment: An Empirical Investigation", unpublished, Stanford University, Department of Economics, December.

Brennan M.J., and E.S. Schwartz (1985): "Evaluating Natural Resource Investments", *Journal of Business* 58: 135–157.

Cukierman A. (1980): "The Effects of Uncertainty on Investment under Risk Neutrality with Endogenous Information", *Journal of Political Economy* 88: 462–475.

Dixit A. (1989): "Entry and Exit Decisions under Uncertainty", *Journal of Political Economy* 97: 620–638.

Evans P. (1984): "The Effects on Output of Money Growth and Interest Rate Volatility in the United States", *Journal of Political Economy* 92: 204–222.

Fischer S., and R.C. Merton (1984): "Macroeconomics and Finance: The Role of the Stock Market", *Carnegie-Rochester Conference Series on Public Policy* 57–108.

Hansen L.P., and K. Singleton (1982): "Generalized Instrumental Variables Estimation of Nonlinear Rational Expectations Models", *Econometrica* 50: 1269–1286.

Ingersoll J.E., Jr., and S.A. Ross (1988): "Waiting to Invest: Investment and Uncertainty", unpublished, Yale University, School of Management, October.

Kester W.C. (1984): "Today's Options for Tomorrow's Growth", *Harvard Business Review*, March/April, 153–160.

Kopcke R.W. (1985): "The Determinants of Investment Spending", *New England Economic Review*, July, 19–35.

Majd S., and R.S. Pindyck (1987): "Time to Build, Option Value, and Investment Decisions", *Journal of Financial Economics* 18: 7–27.

McDonald R., and D.R. Siegel (1986): "The Value of Waiting to Invest", *Quarterly Journal of Economics* 101: 707–728.

Pindyck R.S. (1984): "Risk, Inflation, and the Stock Market", *American Economic Review* 74: 335–351.

Pindyck R.S. (1986): "Capital Risk and Models of Investment Behavior", M.I.T. Sloan School of Management Working Paper No. 1819.

Pindyck R.S. (1988): "Irreversible Investment, Capacity Choice, and the Value of the Firm", *American Economic Review* 78: 969–985.

Pindyck R.S. (1990): "Irreversibility, Uncertainty, and Investment", M.I.T Center for Energy Policy Research Working Paper.

Tatom J.A. (1984): "Interest Rate Variability: Its Link to the Variability of Monetary Growth and Economic Performance", Federal Reserve Bank of St. Louis *Review*, November, 31–47.

Stochastic Models and Option Values
D. Lund and B. Øksendal (Editors)
© Elsevier Science Publishers B.V. (North-Holland), 1991

Financial and Non-financial Option Valuation

Diderik Lund

Department of Economics
University of Oslo
N-0317 Oslo 3, Norway

1. Introduction

The words "option", "option value", and "option price" have been given different meanings in the economics literature. At least three different traditions exist. There has been little communication between the traditions, in particular between the financial tradition on one hand and the non-financial traditions on the other. The purpose of this paper is to bridge the gap between them.

The financial tradition has primarily been concerned with the valuation of the financial securities known as options, e.g. on stocks or on commodities. Uncertainty is central in the model. Excellent surveys are those by Smith (1976) and (1979), and Mason and Merton (1985). The more general concepts of contingent claims or derivative assets have now replaced options as the objects to be valued. The theory, known as contingent claims analysis, can be used to find the market value of various kinds of assets, including real investment projects.

The two other traditions use the word "option" in the more general sense. The focus is on the value of being able to make a choice. In these traditions, too, uncertainty is central to the problem. Following Hanemann (1984), we can distinguish between two traditions: The Schmalensee-Bohm-Graham (SBG) tradition and the Arrow-Fisher-Henry (AFH) tradition. In line with this, I shall call the financial tradition BSM, after the seminal papers of Black and Scholes (1973) and Merton (1973).

In the next section, I shall sketch the relation between these three traditions. Section 3 presents a basic version of the AFH model, and shows how its concept of option value can be interpreted in relation to financial

Thanks to Iulie Aslaksen and Michael Hanemann for comments.

options. Section 4 compares some of the general findings of the AFH and
the BSM models. There is also a general discussion of the applicability
of the two approaches. Section 5 takes up an application from the AFH
literature, and discusses whether the consumer surplus measured there, can
alternatively be measured by BSM techniques.

2. Three traditions: Similarities and dissimilarities

All three traditions are based on some common assumptions, at least in
most of their written work. Consumers/investors are assumed to maximize
expected utility (von Neumann and Morgenstern (1947)), most often in
time-additive form. At each point in time they all have access to the same
information set, and agree on probabilities of future events. In parts of the
financial literature the latter assumption is relaxed.

The non-financial traditions seem to start with Weisbrod (1964), who
gave an informal discussion based on an example: The destruction of a
national park is considered. The calculation of the value of preserving it
should not only take into account its value to those who are expected to
visit it in the future, but also its value to those who are not expected to
do so. Essentially one should calculate the value of having the option to
visit the park.

One interpretation of this is that there is a difference between the
willingness to pay today for preserving the park and the expected consumer
surplus of the park. This is the SBG concept of option value, which I shall
not say much about. It is discussed in Schmalensee (1972), Bohm (1975),
and Graham (1981). It relies on consumers being risk averse. It is not
necessarily positive, as shown by Bohm among others.

Another interpretation is given by Arrow and Fisher (1974). Option
value is interpreted as the value of gaining and/or taking advantage of
more information before making a decision. This value is always positive,
under risk neutrality as well as under risk aversion. Option value can
be identified with the value of flexibility, as opposed to irreversibility. A
decision to destroy the park now is irreversible, whereas the alternative, to
preserve it, is flexible, in the sense that it can be destroyed later.

Arrow and Fisher (1974) called this "quasi-option value", as opposed to
what is called SBG option value above. Henry (1974) was concerned with
the same concept, and called it "the irreversibility effect". The relation
between AFH option value and SBG option value is discussed in Hanemann
(1984).

Models for valuation of financial options date back at least to Bachelier (1900). Starting with Black and Scholes (1973), the most accepted set of assumptions in this literature is quite specific with regard to the stochastic properties of (stock) prices, and to trading opportunities, whereas one makes comparatively weak assumptions on the preferences of consumers/investors, and on their beliefs about the expected drift of prices. What is interesting here is that the value of the financial option over and above the value of comparable securities is really an AFH option value: It derives from the ability to choose in the future whether to exercise the option or not, depending on the stock price which will then be observed. This value is not based on risk aversion.

For some reason the AFH and the BSM traditions only refer to each other in footnotes, see e.g. footnote 7 of Hanemann (1989), footnote 4 of Majd and Pindyck (1987), and footnote 2 of Pindyck (1988). I shall show that the basic concepts are similar, but that the applications to empirical problems have been different. Combining elements of both traditions may be fruitful.

3. The AFH model and the BSM model compared

The following is the simplest AFH model, adopted from Fisher and Hanemann (1986) or (1987). There are two periods, $t = 1, 2$. At the outset the decision maker has the opportunity to make an irreversible investment of a fixed (unit) size, with d_t being the (noncumulative) level of investment undertaken in period t. Because of the irreversibility, there are three possibilities, $(d_1, d_2) = (0,0), = (0,1),$ or $= (1,0)$. The net per-period benefit, as valued at the beginning of period 1, is B_t. B_1 is a function of d_1, while B_2 is a function of d_1 and d_2, and also of a random variable, θ.

An example is the decision to develop a tract of land. The net benefits will depend on the benefit of the undeveloped tract, the benefit of the developed tract, and the investment and operating costs of development. The benefits may be measured in consumption (or money) terms, or in terms of units of the utility function. Below, I shall assume that they are measured in money terms.

If the magnitude of d_2 has to be decided before θ is known, the total net benefit over the two periods can be written as

$$V^*(d_1) = B_1(d_1) + \max_{d_2 \in [0, 1-d_1]} \{E[B_2(d_1, d_2, \theta)]\}. \tag{1}$$

If d_2 can be chosen after θ is known, the total net benefit over the two periods will be

$$\hat{V}(d_1) = B_1(d_1) + E\left\{\max_{d_2 \in [0, 1-d_1]} [B_2(d_1, d_2, \theta)]\right\}. \tag{2}$$

(In (1) and (2) it is assumed that the d_t's can only take the values 0 and 1.) It is clear that $\hat{V}(1) = V^*(1)$, since the choice of $d_1 = 1$ leaves no choice (but $d_2 = 0$) for the second period. The AFH option value is defined as

$$OV = \hat{V}(0) - V^*(0), \tag{3}$$

i.e. the value of getting to know θ before deciding on d_2, conditional on not having made the irreversible decision in period 1, i.e. conditional on $d_1 = 0$.

Denote by d_1^* the value of d_1 which maximizes (1), and by $\hat{d}_1$ the value of d_1 which maximizes (2). Fisher and Hanemann (1986) show that $\hat{d}_1 \leq d_1^*$. The possibility of learning the value of θ before deciding on the investment may cause postponement of the decision, but never the opposite.

The next step is to relate the AFH model to the financial tradition. The procedure will be as follows: First, I state the standard assumptions of financial option valuation models, and a basic result. Second, I show that an American call option on a stock which pays no dividends can be fit into the AFH model, i.e., there exist proper $B_1(d_1)$ and $B_2(d_1, d_2, \theta)$ functions for this case. In this connection I show that the alternative, "starred" scenario can be interpreted as the value of a leveraged stock, and I discuss what the concepts "option value" and "option price" mean in this connection. Third, I discuss what happens if the option is replaced by an option on a stock that does pay dividends.

The standard assumptions in the BSM tradition are: There is continuous trading in securities markets, with no transaction costs, shortselling restrictions, indivisibilities, or taxes. Borrowing or lending at a constant riskless rate, r, is always possible. The stock price, S_t, follows a geometric Brownian motion,

$$\frac{dS_t}{S_t} = \mu dt + \sigma dZ_t, \tag{4}$$

where μ and σ are constants, and Z_t is a standard Wiener process. This implies that the stock price at time 2 is lognormal,

$$S_2 = S_1 \exp\left(\mu - \frac{\sigma^2}{2} + \sigma Z\right), \tag{5}$$

where Z is a standard normal random variable.

A basic result which follows is a method for determining the market value today ($t = 1$) of a claim (e.g., a financial option) to an uncertain cash flow to be received in the future (e.g., at $t = 2$). When the uncertain cash flow is a continuous function of another uncertain variable (the "underlying variable", e.g., the price of a stock, "the underlying asset"), and when claims (e.g., forward contracts on the stock) on this cash flow have a known market value today, the market value of the option-like claim can be determined as follows:

(1) Construct a variable which is perfectly correlated with the underlying variable, but which has a reduced (risk-adjusted) drift term (instead of μ). If the underlying asset has an expected price increase which is sufficient to induce investors to hold it (for the purpose of price gain alone, account taken of the risk), the reduced drift term should be the riskless rate, r. The asset price may, however, have a rate-of-return shortfall, which I shall assume to be constant, δ (see McDonald and Siegel (1984)). (For a stock, this may result from a constant proportional dividend yield, as in one of the models of Merton (1973). For a commodity, it can be interpreted as a constant proportional marginal net convenience yield, as in Brennan and Schwartz (1985).) In this case the drift term should be $r - \delta$.

(2) Calculate the expected value of the uncertain (option-like) cash flow, conditional on this risk-adjusted process of the underlying asset. The market value today is found by discounting this expected value at the riskless rate.

The method is well-known for linear functions of underlying variables, for which it reduces to using a risk-adjusted discount rate. For non-linear functions, the basic result is that the risk-adjustment should take place in the drift term of the underlying asset, not in the discount rate.

The method was introduced by Cox and Ross (1976), who showed that it holds for European call options. Since any continuous function of underlying assets can be approximated arbitrarily closely by linear combinations of such options, it holds for all continuous functions. Constantinides (1978) and Cox, Ingersoll, and Ross (1985) derive the same result in general equilibrium models. These do not require that the underlying variables are actually traded assets, but they make stronger assumptions in order to characterize financial market equilibrium.

Return now to a particular kind of option. Consider an American call option on a stock which is known not to pay dividends between today and the option's expiration date. A well-known result in the BSM tradition is that such an option should never be exercised before its expiration date (Merton (1973)). I want to examine this option's value within the AFH model.

Let S_1 be today's stock price. The decision $d_t = 1$ is the decision to liquidate one's portfolio at time t, while $d_t = 0$ is the decision to hold on to the portfolio after time t. The financial models most often operate in continuous time, or with many periods. Assume for the sake of comparison that there are only two periods, $t = 1, 2$, as in the AFH model above.

The AFH model compares two different "scenarios", i.e., two different assumptions with regard to the sequence of information arrival and investment decision. For the option, information on the stock price at $t = 2$, S_2, arrives before one has to decide on exercising it. S_2 is the random variable that takes the place of θ. Considering the definition (2), this means that the essential feature of $\hat{V}(0)$ is captured. It is the "hatted" scenario that describes the financial option.

I shall show that the alternative, "starred" scenario can also be interpreted as giving the value of a portfolio of financial securities, namely a leveraged stock position. This consists of one share of the stock and a loan, i.e., a short sale of bonds, with the value today equal to the option's exercise price, K. I shall return to a slight problem with this interpretation.

I shall first define the functions $B_1(d_1)$ and $B_2(d_1, d_2, \theta)$ so that $\hat{V}(d_1)$ from (2) will be equal to the BSM market value of the option. If exercised today, the option has the value $S_1 - K$. The function $B_1(d_1)$ should thus be defined by $B_1(0) = 0$, $B_1(1) = S_1 - K$. One gets $\hat{V}(1) = B_1(1) = S_1 - K$. Of course, $B_2(1, 0, S_2) = 0$.

The existence of a suitable B_2 function may not be so obvious, but is actually in order. Under the standard BSM assumptions one can use the Cox and Ross (1976) valuation method, in which the market value of the option is really an expected value of a maximum. The B_2 function for this option will be denoted B_2^a. It must satisfy $B_2^a(0, 1, S_2) = (S_2^a - K) \exp(-r)$, where S_2^a is a risk-adjusted process for the price S_2. S_2^a is perfectly correlated with S_2, but has a risk-adjusted drift term, so that $E(S_2^a) = S_1 \exp(r)$. (If the stock could pay dividends, $E(S_2^a)$ would be lower.) The definition of S_2^a is thus $S_2^a \equiv S_2 \exp(r - \mu)$. Of course, $B_2^a(0, 0, S_2) \equiv 0$.

A natural question to ask is whether all kinds of option-like opportunities over two periods can be described by the equation (2). Can the value of holding on to the option until new information arrives always be written as the expected value of a maximum? I believe that this follows directly from the expected utility formulation. For marginal changes in utility it will be true even for non-time-additive utility functions. It will be true when benefits are measured in units of the utility function. But after division by the marginal utility of the consumption good in each state, it will also be true when benefits are measured in consumption terms.

This does not mean that the Cox and Ross method is a trivial result that can be derived from the expected utility formulation. That formulation tells us that for each agent, the benefit at $t = 2$ can be written as the expected value of the maximum of some function B_2, but not that the B_2^a function is as simple as it comes out, and the same for everyone. From the Black and Scholes formula alone (see equation (8) below), this is not transparent.

Having determined what the $B_2^a(0, d_2, S_2)$ function has to be, the B functions should now be carried over to the "starred" scenario. For $d_1 = 1$, one gets $V^*(1) = B_1(1) = S_1 - K$. Consider then the case of $d_1 = 0$. The value at $t = 1$ of deciding at $t = 1$ to exercise the option at $t = 2$ is, according to the definition of B_2^a above,

$$E[B_2^a(0, 1, S_2)] = E(S_2^a - K)\exp(-r) = S_1 K \exp(-r). \qquad (6)$$

The alternative is to decide on not exercising, which has the value of $B_2^a(0, 0, S_2) = 0$. If the decision had to be made at $t = 1$, the choice would depend on whether $S_1 \gtrless K \exp(-r)$, and the value is $V^*(0) = \max[0, S_1 - K \exp(-r)]$.

It provides intuition to try to find a portfolio of securities which has the characteristics of the "starred" scenario. A leveraged stock position, with a debt of K, has the value $S_1 - K$ if liquidated at $t = 1$.

Consider now the case $d_1 = 0$. For this portfolio and a planning horizon of two periods, there does not seem to be any natural way to find the analogy of making a choice at $t = 1$ whether to liquidate at $t = 2$ or not. I shall therefore confine myself to looking for an analogy when $S_1 > K \exp(-r)$, in which case the "starred" scenario implies a decision $d_2 = 1$. The value at $t = 1$ of deciding at $t = 1$ to liquidate the position at $t = 2$ is

$$E[S_2^a - K \exp(r)]\exp(-r) = S_1 - K. \qquad (7)$$

This shows that the analogy is incomplete even when $S_1 > K \exp(-r)$. The difference between (6) and (7) is due to the fact that the option's exercise price does not accumulate at the rate of interest. Whether exercised at $t = 1$ or at $t = 2$, the exercise price is K. If we had started out with a non-standard call option with an exercise price at $t = 2$ of $K \exp(r)$, this would have affected the definition of $B_2^a(0, 1, S_2)$. Then the analogy between the "starred" scenario and the leveraged stock postition would have been exact as long as $S_1 > K$, i.e., as long as the option is "in the money" at $t = 1$.

The four values of the V functions can be summarized:

$$\hat{V}(1) = V^*(1) = S_1 - K, \tag{8}$$
$$V^*(0) = \max[0, S_1 - K \exp(-r)],$$
$$\hat{V}(0) = E\{\max[0, S_2 \exp(r - \mu) - K] \exp(-r)\}$$
$$= E\{\max[0, S_2 \exp(-\mu) - K \exp(-r)]\}$$
$$= S_1 N(x_0) - K N(x_0 - \sigma) \exp(-r),$$

where

$$x_0 = \frac{[\log(S_1/K) + r]}{\sigma} + \frac{\sigma}{2}.$$

The last expression for $\hat{V}(0)$ is of course the one-period version of the formula of Black and Scholes (1973), N being the standard cumulative normal distribution function. An explicit, direct derivation of the fifth line of (8) from the fourth is given in Jarrow and Rudd (1983).

The AFH "option value" from (3) can now be found for an American call option on a stock with no dividends. It is simply the market value $\hat{V}(0)$ at $t = 1$ of the option, if kept alive (not exercised), minus the value $V^*(0) = \max[0, S_1 K \exp(-r)]$. There may be some confusion as the AFH tradition calls $OV = \hat{V}(0) - V^*(0)$ the "option value", whereas for this kind of options, the BSM tradition calls $\hat{V}(0)$ the "option value". (More generally the BSM tradition would use "option value" for $\max[\hat{V}(1), \hat{V}(0)]$, which in this case is always $\hat{V}(0)$.) In figure 1, the BSM option value is the convex curve, while the AFH option value is the difference between this curve and the kinked line, $V^*(0)$.

The central result that OV is positive is shown e.g. in Hanemann (1989). The same result is known in the BSM tradition: Merton (1973) shows that the value of a (European or American) call option on a stock that pays no dividends always exceeds $\max[0, S_1 - K \exp(-r)]$. This result is shown under much weaker conditions than the standard set listed above, with no restriction on the stock price movement except limited liability.

In the BSM tradition, the term "option price" refers to the observed market price, and one assumes that this normally coincides with the "option

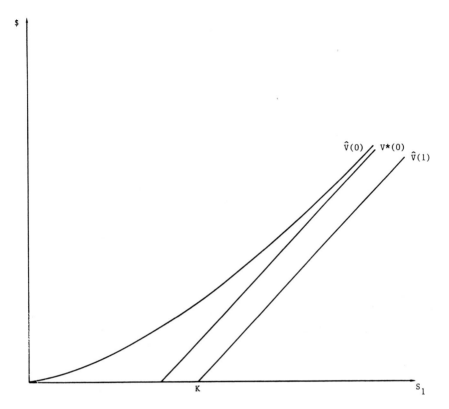

Figure 1: Black and Scholes option value as a function of the stock price at $t = 1$, and its relation to AFH option value: The case of a stock that pays no dividends. The option expires at $t = 2$.

value". (If not, one says that the option is "overpriced" or "underpriced".) Hanemann (1989) writes that "in terms of the financial literature, $\hat{V}(0)$ is the value of the "option", and $\hat{V}(1)$ is its "price" (opportunity cost)", (footnote 7). I find this misleading. In the BSM tradition, the magnitude $\hat{V}(1)$ is known as the value of the option if it is not kept alive, i.e., if it is exercised. For the case I have discussed here, it is not optimal to exercise the option before its expiration date. It is worth more alive than dead, $\hat{V}(0) > \hat{V}(1)$.

   This leads directly to a discussion of the case of an American call option on a stock which pays dividends. Consider the case when the stock pays a continuous dividend proportional to the stock price, $D_t = \delta S_t$. Merton

(1973, footnote 62) shows that the option may be worth more dead than alive, i.e., that it is optimal to exercise the option prematurely if the stock price is high. He shows that the value of the corresponding European option is bounded below by $S_1 \exp(-\delta) - K \exp(-r)$ as $S_1 \to \infty$. The optimale exercise strategy, and the value of the option, are analyzed by Samuelson (1965), and in a more general setting by McDonald and Siegel (1986).

This case does not fit nicely into the two-period AFH model. The reason is that the BSM models assume that an American call option can be exercised at any moment in continuous time, and in this case, it may actually be optimal to utilize this opportunity. An artificial semi-American option that would fit into the AFH model would keep only two choices open: Exercise at $t = 1$ or at $t = 2$. This kind of option would reduce to a European option if the owner decides not to exercise at $t = 1$. This non-standard option value can be analyzed following the scheme above.

The BSM tradition would in this case use the term "option value" for $\max[\hat{V}(1), \hat{V}(0)]$, which is now different from $\hat{V}(0)$ for large values of S_1. The case is illustrated in figure 2, for $r < \delta$. It is optimal to exercise the option at $t = 1$ if $S_1 > \hat{S}_1$. The B_2 function for the case of $d_1 = 0$ is still zero for $d_2 = 0$. For $d_2 = 1$, one can still use the Cox and Ross method, but the risk-adjusted stock price is now $S_2^d \equiv S_2 \exp(r - \mu - \delta)$. Define $B_2^d(0, 1, S_2) \equiv (S_2^d - K) \exp(-r)$. The four different values of the V functions are now,

$$\hat{V}(1) = V^*(1) = S_1 - K, \tag{9}$$
$$V^*(0) = \max[0, S_1 \exp(-\delta) - K \exp(-r)],$$
$$\hat{V}(0) = E\{\max[0, S_2 \exp(r - \mu - \delta) - K] \exp(-r)\}$$
$$= E\{\max[0, S_2 \exp(-\mu - \delta) - K \exp(-r)]\}$$
$$= S_1 N(x) \exp(-\delta) - K N(x - \sigma) \exp(-r),$$

where

$$x = \frac{[\log(S_1/K) + r - \delta]}{\sigma} + \frac{\sigma}{2},$$

see, e.g., Smith (1976) or McDonald and Siegel (1984).

The convex curve $\hat{V}(0)$ in figure 2 is the value of a European call option with one period left to maturity. For a standard American call option on this stock, the value would be higher, since it gives the owner the additional opportunity to exercise the option at any moment between $t = 1$ and $t = 2$.

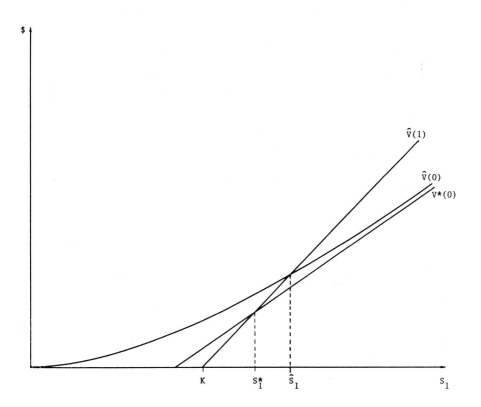

Figure 2: Black and Scholes option value as a function of the stock price at $t = 1$, and its relation to AFH option value: The case of a stock that pays continuous, proportional dividends. The option can be exercised at $t = 1$ or at $t = 2$.

4. The AFH and BSM models: Results and applicability

The preceding section showed that financial option valuation can be considered as one application of the AFH option (or "quasi-option") model. The main limitation is that the basic AFH model refers to only two points in time. A generalization of this can be found in Fisher and Hanemann (1985).

The AFH model is more general — in fact it is a rather abstract framework which illustrates the concept of the value of flexibility. Even within this abstract framework it has been possible to demonstrate important results: Neglecting future flexibility, i.e., basing decisions today merely on expected future cash flows, may lead to too early investment.

A second result of the AFH model is that option value increases with uncertainty. In Theorem 1 of Hanemann (1989), an increase in uncertainty is specified as a mean-preserving spread (MPS) in the distribution of θ, and it is shown that when B_2 is linear in θ, an MPS will lower $\hat{d}_1$, will have no effect on d_1^*, and will raise OV. (The B_2 functions of the previous section are really linear in S_2 in each case.) The same results are well-known from the financial models: Increased uncertainty increases option value, (Merton (1973), Theorem 8), and postpones investment (McDonald and Siegel (1986)). That d_1^* is unaffected is clear from the risk-neutrality argument of Cox and Ross (1976).

It does not seem that the general results of the AFH and the BSM traditions are very different. The general insights are the same in both camps. The rest of this section will discuss how one might choose which method to adopt, concluding that the methods are applicable in different situations. I shall also argue that there are other reasons why researchers should be aware of both schools, namely that each provides some insights that can hardly be obtained within the other's framework.

For any application, the choice between the AFH and the BSM models must depend on whether the more restrictive assumptions of the latter are fulfilled. For two-period problems the BSM model is a special case of the AFH model, and adding assumptions adds a lot to the precision of the answers one gets. The dillemma is well known in economics: Is it possible to use market data (prices of commodities and financial assets) to calculate the values of decision criteria, that are derived from normative models? Can one infer from the market's valuation of something what value the decision maker should put on similar things?

The BSM model makes some fairly strong assumptions about the stochastic properties of prices of the underlying assets. The justification of the assumptions are given by Merton (1982), although that paper does not spell out the exact connection to the underlying assumption of efficient financial markets. There is a problem, however, to maintain these assumptions when the BSM model is applied to the valuation of "real options", as has become quite popular (see Mason and Merton (1985) for references). For the spot prices of (extracted units of) exhaustible resources, there exists a theoretical justification for an assumption that relative price changes have no serial correlation. For other commodity prices, the assumption seems more far-fetched.

In many cases, it may be advantageous first to derive results under the weak assumptions of the AFH model, and then to derive special results under the assumptions of the BSM model. An example from Norway is a discussion in 1988 on whether to develop a particular oil field now, or to postpone the decision. The academics who advocated the postponement relied on the BSM assumptions. The assumption of the oil price following a lognormal diffusion process was criticized by their opponents as unsubstantiated. However, the advocates for immediate development basically showed that this was profitable when flexibility was ignored. It should be possible to argue, in line with the AFH model, that ignoring flexibility leads to too early development, regardless of the assumptions on oil price uncertainty. An exact number for the loss of disregarding flexibility requires estimation of the B_2 function, and for this purpose, it is tempting to try the BSM model and use some market data.

For many applications, those market data are not available. The AFH tradition has been particularly concerned with the valuation of the environment, and in particular the option value in this connection. Since environmental qualities are rarely traded in markets, and may be of a public-good character, it is not likely that the BSM model can add much to these analyses.

A particular advantage of the BSM model is that it is consistent with the lessons from finance theory about the valuation of risk in well-developed financial markets. It is thus consistent with agents being risk-averse, without having to specify the degree of their risk aversion. This will often remain a problem in other AFH applications.

Finally, I shall point out that the AFH model is general enough to include many related concepts that I have not discussed so far. Hanemann (1989) shows the relation to the concept of expected value of perfect information (EVPI) as defined by Conrad (1980), and to the option value concept of Bernanke (1983). These clarifications are important, since none of these authors in fact deal with the same concept as the original AFH papers did. As mentioned above, Hanemann (1984) also draws the line to the SBG option value.

5. An example of a hybrid method: Uncertain consumer surplus

Fisher and Hanemann (1986) gives an example of how to estimate AFH option value for a non-marginal change in the supply curve of corn. For non-marginal changes, consumer and producer surplus considerations are necessary to estimate net benefits. In this section I shall combine this with financial valuation methods. The example below is not intended as a realistic case. It is meant to demonstrate that the two traditions may be seen in conjunction. It is a first attempt to use the BSM methods, based on assumptions about financial markets, in a non-standard way, inspired by the AFH tradition.

Consider an economy in which there is a non-stochastic downward sloping demand curve (linear for simplicity) for some commodity at $t = 2$, $Q(S_2)$. As viewed from $t = 1$, S_2 is uncertain, as given by (5). The price S_t has a constant rate-of-return shortfall of δ. This means that the forward price at $t = 1$ of one unit to be delivered at $t = 2$ is $S_1 \exp(r - \delta)$. For simplicity one may assume that such forward contracts are traded, although it is sufficient that the financial markets are sufficiently well developed that such contracts can be constructed from traded assets.

Only the consumer surplus will be considered here. This may be interpreted as an assumption of constant marginal costs, or as an assumption that the economy is small and open, a price taker internationally, and does not care about the producer surplus of the imported commodity.

Figure 3 illustrates the consumer surplus, when $Q(S_2) = \alpha - \beta S_2$. The question to be analyzed here is: What is the value at $t = 1$ of having the option to buy (or produce at constant marginal costs) the commodity at $t = 2$ at a non-stochastic price of K? This is a non-marginal option, valid for all units of the commodity to be bought at $t = 2$. Thus it affects the whole consumer surplus, the shaded area in the figure.

The gross benefits from buying the commodity are given by the non-stochastic demand curve, and the valuation of these is unproblematic. They should be discounted by the riskless rate. The costs are also valued equally by all consumers, by the assumption of the properties of financial markets. Everyone has the same marginal valuation of (i.e., willingness to pay for avoiding) this uncertain cost, but this marginal valuation is constant for all units of the commodity.

This rests on an assumption of a non-changing risk adjustment in the valuation of the costs. If the costs have a probability distribution across states of the world which is unique in the economy, the risk adjustment would be altered by a non-marginal change in the quantity of the commodity. Assume instead that the costs are spanned by traded assets, and are small in relation to the economy, so that the risk adjustment does not change.

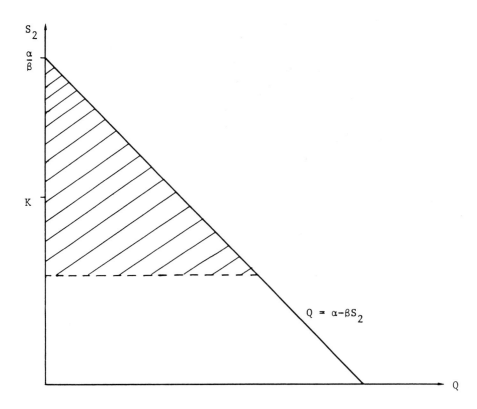

Figure 3: Consumer surplus (the shaded area) as a function of the price
at $t = 2$, S_2, when the demand function is linear.

The consumer surplus, the shaded area, is

$$CS(S_2) = \begin{cases} \alpha^2/2\beta - \alpha S_2 + (\beta/2)(S_2)^2 & \text{for } S_2 \leq \alpha/\beta, \\ 0 & \text{for } S_2 > \alpha/\beta. \end{cases} \quad (10)$$

Figure 4 illustrates. The graph of CS is the downward sloping part of the
parabola, and extends to the right along the horizontal axis.

Consider first the value at $t = 1$ of receiving this uncertain magnitude
at $t = 2$, with no option. Since it is a continuous function of S_2, it can be
valued by the Cox and Ross (1976) method. S_2 should be replaced by S_2^d,
and one calculates the expected, discounted value,

$$E[CS(S_2^d)]\exp(-r) \quad (11)$$

$$
= E\left[\frac{\alpha^2}{2\beta} - \alpha S_2^d + \left(\frac{\beta}{2}\right)(S_2^d)^2 \Big| S_2^d \le \frac{\alpha}{\beta}\right] \Pr\left(S_2^d \le \frac{\alpha}{\beta}\right)\exp(-r)
$$

$$
= \left(\frac{\alpha^2}{\beta}\right)\left\{1 - N\left[x\left(\frac{\alpha}{\beta}\right) - \sigma\right]\right\}\exp(-r)
$$

$$
- \alpha S_1 \exp(-\delta)\left\{1 - N\left[x\left(\frac{\alpha}{\beta}\right)\right]\right\}
$$

$$
+ \left(\frac{\beta}{2}\right)(S_1)^2 \exp(r - 2\delta + \sigma^2)\left\{1 - N\left[x\left(\frac{\alpha}{\beta}\right) + \sigma\right]\right\},
$$

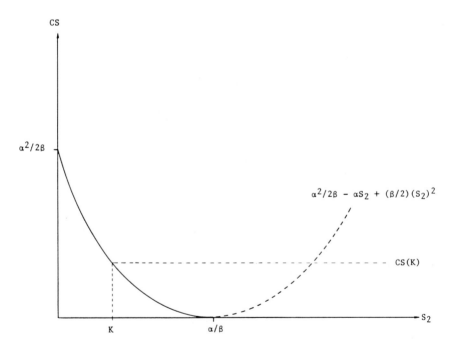

Figure 4: Consumer surplus as a function of S_2, with and without an upper bound, K, on unit costs.

as shown in the appendix. The function $x(.)$ is defined by

$$x(K) \equiv \frac{\left[\log\left(\frac{S_1}{K}\right) + r - \delta\right]}{\sigma} + \frac{\sigma}{2}. \tag{12}$$

If one includes the option to buy at a nonstochastic price K, the surplus at $t = 2$ will be $\max[CS(K), CS(S_2)]$. This is also a continuous function of S_2, and its graph is shown in figure 4: It includes the upper section of the downward sloping part of the parabola, and then extends along the dashed, horizontal line to the right.

The market value of this function is, using the appendix,

$$E\{\max[CS(K), CS(S_2^d)]\} \exp(-r) \tag{13}$$
$$= \left\{ \frac{\alpha^2}{2\beta} + \left[-\alpha K + \left(\frac{\beta}{2}\right) K^2\right] \Pr\left(S_2^d > K\right) \right.$$
$$\left. + E\left[-\alpha S_2^d + \left(\frac{\beta}{2}\right)(S_2^d)^2 | S_2^d \leq K\right] \Pr\left(S_2^d \leq K\right) \right\} \exp(-r)$$
$$= \frac{\alpha^2}{2\beta} \exp(-r) + \left[-\alpha K + \left(\frac{\beta}{2}\right) K^2\right] \exp(-r) N[x(K) - \sigma]$$
$$- \alpha S_1 \exp(-\delta)\{1 - N[x(K)]\}$$
$$+ \left(\frac{\beta}{2}\right)(S_1)^2 \exp(r - 2\delta + \sigma^2)\{1 - N[x(K) + \sigma]\}.$$

Of course, (13) reduces to (11) if the upper bound on the cost, K, is equal to α/β.

A natural extension will be to investigate whether the expression (13), the expression (11), or their difference, are increasing functions of risk. Two different concepts of increasing risk could be employed, either a mean-preserving spread in S_2 (following Hanemann (1989)), or an increase in σ.

For the second of these possibilities, one should observe that Merton (1973) shows in an appendix that the expected value of a concave function of a lognormal variable (such as S_2^d) is a decreasing function of σ. This implies that expected values of convex functions (such as (11) and (13)) of lognormal variables are increasing functions of σ.

The conclusion is that it has been possible to extend the financial valuation of options to include consumer surplus considerations. More detailed investigations of the resulting formulas should be carried out.

Appendix

The appendix gives the calculations leading to equations (11) and (13). The calculations are straightforward, by an extension of the calculations in Jarrow and Rudd (1983). The same method will work for the expectation of any power function of a lognormal variable, which means that other demand functions which are power functions will also give explicit answers.

First, observe that from (5) and the definition of S_2^d,

$$S_2^d \equiv S_1 \exp\left(r - \delta - \frac{\sigma^2}{2} + \sigma Z\right), \qquad (A.1)$$

where Z is standard normal. The probability that this variable exceeds a number K is,

$$\Pr\left(S_2^d > K\right) = \Pr\left\{Z > \frac{\left[\log\left(\frac{K}{S_1}\right) - r + \delta + \frac{\sigma^2}{2}\right]}{\sigma}\right\} \qquad (A.2)$$
$$= N[x(K) - \sigma],$$

using the definition of $x(K)$ in equation (12).

In addition, I need two conditional expectations, multiplied by the corresponding probability. The first one is,

$$E(S_2^d | S_2^d \leq K)\Pr\left(S_2^d \leq K\right) = \int_0^K S_2^d f(S_2^d) dS_2^d, \qquad (A.3)$$

where $f(.)$ is the density function of S_2^d. Observe that $S_2^d \to 0^+$ implies $Z \to -\infty$, and that $S_2^d = K$ implies $Z = \sigma - x(K)$. The expression in (A.3) thus becomes

$$\int_{\infty}^{\sigma - x(K)} S_1 \exp\left[\left(r - \delta - \frac{\sigma^2}{2}\right) + \sigma z\right] \exp\left(-\frac{z^2}{2}\right)(2\pi)^{-1/2} dz \quad (A.4)$$
$$= S_1 \exp(r - \delta) \int_{x(K)}^{\infty} \exp\left(-\frac{w^2}{2}\right)(2\pi)^{-1/2} dw$$
$$= S_1 \exp(r - \delta)\{1 - N[x(K)]\},$$

where the substitution $w \equiv \sigma - z$ has been used.

Similarly,

$$E\left[(S_2^d)^2|S_2^d \le K\right]\Pr\left(S_2^d \le K\right) = \tag{A.5}$$

$$\int_{|\infty}^{\sigma-x(K)}\left\{(S_1)^2\exp\left[2\left(r-\delta-\frac{\sigma^2}{2}\right)+2\sigma z\right]\right\}\exp\left(-\frac{z^2}{2}\right)(2\pi)^{-1/2}\,dz$$

$$=(S_1)^2\exp[2(r-\delta)+\sigma^2]\{1-N[x(K)+\sigma]\}\,,$$

where $v \equiv 2\sigma - z$ is a suitable substitution.

References

Arrow K.J., and A.C. Fisher (1974): "Environmental Preservation, Uncertainty, and Irreversibility", *Quarterly Journal of Economics* 88: 312–319.

Bachelier L. (1900): "Theory of Speculation", English translation in P. Cootner (editor) (1964), *The Random Character of Stock Market Prices,* Cambridge, Mass.: MIT Press, 17–78.

Bernanke B.S. (1983): "Irreversibility, Uncertainty, and Cyclical Investment", *Quarterly Journal of Economics* 98: 85–106.

Black F., and M. Scholes (1973): "The Pricing of Options and Corporate Liabilities", *Journal of Political Economy* 81: 637–659.

Bohm P. (1975): "Option Demand and Consumer's Surplus: Comment", *American Economic Review* 65: 733–736.

Brennan M.J., and E.S. Schwartz (1985): "Evaluating Natural Resource Investments", *Journal of Business,* 58: 135–157.

Conrad J.M. (1980): "Quasi-Option Value and the Expected Value of Information", *Quarterly Journal of Economics* 94: 813–820.

Constantinides G.M. (1978): "Market Risk Adjustment in Project Valuation", *Journal of Finance* 33(2): 603–616.

Cox J.C., J.E. Ingersoll, jr., and S.A. Ross (1985): "An Intertemporal General Equilibrium Model of Asset Prices", *Econometrica* 53(2): 363–384.

Cox J.C., and S.A. Ross (1976): "The Valuation of Options for Alternative Stochastic Processes", *Journal of Financial Economics* 3: 145–166.

Fisher A.C., and W.M. Hanemann (1985): "Valuing Pollution Control: The Hysteresis Phenomenon in Aquatic Ecosystems", Working Paper No. 361, University of California, Department of Agricultural and Resource Economics, Berkeley, August.

Fisher A.C., and W.M. Hanemann (1986): "Option Value and the Extinction of Species", in V.K. Smith (editor), *Advances in Applied Micro-Economics: Risk Uncertainty and the Valuation of Benefits and Costs,* Vol. 4, Greenwich, Conn.: JAI Press, 169–190.

Fisher A.C., and W.M. Hanemann (1987): "Quasi-Option Value: Some Misconceptions Dispelled", *Journal of Environmental Economics and Management* 14: 183–190.

Graham D.A. (1981): "Cost-Benefit Analysis Under Uncertainty", *American Economic Review* 71: 715–725.

Hanemann W.M. (1984): "On Reconciling Different Concepts of Option Value", University of California, Division of Agricultural Sciences, Berkeley, July.

Hanemann W.M. (1989): "Information and the Concept of Option Value", *Journal of Environmental Economics and Management* 16: 23–37.

Henry C. (1974): "Investment Decisions Under Uncertainty: The "Irreversibility Effect"," *American Economic Review* 64(6): 1006–1012.

Jarrow R.A., and A. Rudd (1983): *Option Pricing,* Homewood, Ill.: Dow Jones Irwin.

Majd S., and R.S. Pindyck (1987): "Time to Build, Option Value, and Investment Decisions", *Journal of Financial Economics* 18: 7–27.

Mason S.P., and R.C. Merton (1985): "The Role of Contingent Claims Analysis in Corporate Finance", in E.I. Altman and M.G. Subrahmanyam (editors), *Recent Advances in Corporate Finance,* Homewood, Ill.: Irwin.

McDonald R.L., and D.R. Siegel (1984): "Option Pricing When the Underlying Asset Earns a Below-Equilibrium Rate of Return: A Note", *Journal of Finance* 39: 261–264.

McDonald R.L., and D.R. Siegel (1986): "The Value of Waiting to Invest", *Quarterly Journal of Economics* 101: 707–727.

Merton R.C. (1973): "Theory of Rational Option Pricing", *Bell Journal of Economics and Management Science* 4(1): 141–183.

Merton R.C. (1982): "On the Mathematics and Economics Assumptions of Continuous-Time Models", in W.F. Sharpe (editor), *Financial Economics: Essays in Honor of Paul Cootner,* Amsterdam: North-Holland, 19–51.

Neumann J. von, and O. Morgenstern (1947): *Theory of Games and Economic Behavior,* second edition, Princeton, New Jersey: Princeton University Press.

Pindyck R.S. (1988): "Irreversible Investment, Capacity Choice, and the Value of the Firm", *American Economic Review* 78(2): 969–985.

Samuelson P.A. (1965): "Rational Theory of Warrant Pricing", *Industrial Management Review* 6: 13–31.

Schmalensee R. (1972): "Option Demand and Consumer's Surplus: Valuing Price Changes under Uncertainty", *American Economic Review* 62: 814–824.

Smith C.W., jr. (1976): "Option Pricing, A Review", *Journal of Financial Economics* 3: 3–51.

Smith C.W., jr. (1979): "Applications of Option Pricing Analysis", in J.L. Bicksler (editor), *Handbook of Financial Economics,* Amsterdam: North-Holland, 79–121.

Weisbrod B.A. (1964): "Collective-Consumtion Services of Individualized-Consumption Goods", *Quarterly Journal of Economics* 78: 471–477.

III. Stochastic Control and
Dynamic Programming

Stochastic Models and Option Values
D. Lund and B. Øksendal (Editors)
Elsevier Science Publishers B.V. (North-Holland), 1991

Partial Investment Under Uncertainty

T.Ø. Kobila[*][†]

1. Introduction

The importance of irreversibility in investment decisions is currently receiving renewed attention in the literature. Recently developed stochastic optimization techniques offer new insight into the combined effect of uncertainty and irreversibility. Important and diverse applications include Brennan and Schwartz (1985), McDonald and Siegel (1986) and Pindyck (1988). Pindyck (1988) solves the problem of optimal capacity choice and capacity expansion under uncertainty in future demand and irreversibility of investment. He explicitly states the problem of evaluating a marginal unit of capacity as an option value problem and establishes the link to financial option value techniques.

In this paper we analyze a general class of investment problems under uncertainty and irreversibility. We consider the optimal investment in irreversible capacity for a profit-maximizing firm with the profit function subject to random fluctuations. This general representation of uncertainty includes the cases of demand uncertainty, stochastic product price, input price uncertainty or random disturbances in the production function or cost function. The term "partial investment" is coined to describe the optimal investment strategy where the firm incrementally expands capacity, in contrast to lumpy investments where the entire production capacity is installed at once.

In Section 2 we specify the uncertainty as a geometric Brownian motion and derive a generalized Hamilton-Jacobi-Bellman equation to characterize the optimal investment decision. The optimal investment rule turns out to

---

[*] This is an acronym for Iulie Aslaksen, Olav Bjerkholt and Kjell Arne Brekke, all at the Central Bureau of Statistics, Box 8131 Dep, N-0033 Oslo, Norway, and Tom Lindstrøm and Bernt Øksendal, both at the Institute of Mathematics, University of Oslo, Box 1053 Blindern, N-0316 Oslo 3, Norway.

[†] *Acknowledgement:* The work of B. Øksendal was partially supported by NAVF, Norway. Kobila thanks Robert McDonald for comments to the first version of this paper.

be singular, with the rate of capacity expansion equal to zero or infinity depending on the random fluctuations and the level of capacity already installed.

In Section 3 we discuss the results of Section 2 in the light of neoclassical investment theory. We find that the optimal capacity is smaller when uncertainty and irreversibility are taken into account. This is in accord with the general results obtained under certainty in Arrow (1968) and Nickell (1978). The conclusion that optimal capacity is lower depends critically on the assumption of irreversible investment. When investment is reversible, uncertainty in itself may lead to an increase in optimal capacity, see e.g. Abel (1983). Abel's argument is analogous to that of option value theory, in that optimal capacity is higher, just as a call option on a stock is worth more, the more volatile is the price of the stock. Combined with irreversibility, however, the effect of uncertainty is to lower optimal investment. Finally, we discuss the relationship between our stochastic control model and an option value model, and show that the optimal investment strategy of Pindyck (1988) is a special case of our model.

2. The model

Consider a firm with profit depending on a stochastic parameter Θ_t and capital K_t. Let $\pi(\theta, k)$ denote the profit function, exclusive of capacity acquisition, and suppose that the returns to capacity are increasing in θ and decreasing in k, i.e. $\partial^2 \pi / \partial \theta \partial k \geq 0$ and $\partial^2 \pi / \partial k^2 \leq 0$. Assume that capacity can be increased instantaneously, and that the capacity expansion is irreversible. Assume furthermore that the cost of capacity expansion $C(K_t)$ is a function of existing capacity, K_t, with $C'(K_t) > 0, C''(K_t) \geq 0$. The optimization problem of the firm is then:

$$H(t, \theta, k) = \sup_{u_s} E^{t, \theta, k} \left[\int_t^\infty (\pi(\Theta_s, K_s) - C(K_s)u_s)e^{-rs} ds \right], \qquad (1)$$

where u_s is the rate of capacity expansion such that $dK_s = u_s ds$. Assume furthermore that Θ_t is an Itô diffusion,

$$d\Theta_t = a(\Theta_t)dt + b(\Theta_t)dB_t, \qquad (2)$$

where B_t is a Brownian motion.

The solution to (1) turns out to be of a singular type with a "forbidden region" where the process cannot stay for any positive amount of time.

The usual sufficient condition for optimality that H is C^2 and solves the Hamilton-Jacobi-Bellman (HJB) equation turns out to be too restrictive.

We will derive a modified optimality condition in the form of a generalized HJB-equation. First, we will briefly outline the structure of the solution. Since there is no upper bound on the size of investment, optimal investment will either be 0 or ∞. Under reasonable assumptions we cannot have $u_t = \infty$ on a time interval of positive length, hence, the solution will be of the following form: There exists an open subset $\mathcal{A}$ of the (θ, k)-plane such that $u_t = \infty$ if and only if $(\Theta_t, K_t) \in \mathcal{A}$. If the process starts in $\mathcal{A}$, it cannot stay there for a time interval of positive length, hence, it will immediately be thrown out of this area. Similarly, it is impossible for the process to ever enter the interior of this area, the process will thus "live" outside $\mathcal{A}$.

It turns out that a sufficient condition for optimality is that H solves the generalized Hamilton-Jacobi-Bellman equation. To formulate this generalization we define the operator:

$$\mathcal{L}^u h = u\frac{\partial h}{\partial k} + \frac{\partial h}{\partial t} + a(\theta)\frac{\partial h}{\partial \theta} + \frac{1}{2}b^2(\theta)\frac{\partial^2 h}{\partial \theta^2}. \tag{3}$$

The traditional HJB-equation states that if h satisfies the equation:

$$\sup_{v \geq 0}\left[\mathcal{L}^v h + (\pi(\theta, k) - C(k)v)e^{-rt}\right] = 0, \tag{4}$$

then $h = H$, where H is the optimal value function defined in (1). This can be generalized to the following theorem, where the relation between the region $\mathcal{A}$ and the function $k = \phi(\theta)$ is given by

$$\mathcal{A} = \{(\theta, k) : k < \phi(\theta), \quad \theta \geq 0\}. \tag{5}$$

Theorem 1. Suppose there exists a bounded continuous function $k = \phi(\theta)$ and a function $h(t, \theta, k)$ which is C^1 in t and k, and C^2 in θ, and such that the following conditions are fulfilled:

$$\sup_{v \geq 0}\left[\mathcal{L}^v h(t, \theta, k) + (\pi(\theta, k) - vC(k))e^{-rt}\right]\begin{cases} = 0 & \text{if } k \geq \phi(\theta), \\ \leq 0 & \text{if } k < \phi(\theta), \end{cases} \tag{6}$$

and

$$\frac{\partial h}{\partial k}(t, \theta, k) - C(k)e^{-rt}\begin{cases} < 0 & \text{if } k > \phi(\theta), \\ = 0 & \text{if } k \leq \phi(\theta). \end{cases} \tag{7}$$

Moreover, suppose that there exists $M < \infty$ such that for all $v \geq 0$, $t \geq 0$ and $k < \phi(\theta)$

$$\mathcal{L}^v h(t, \theta, k) + (\pi(\theta, k) - vC(k))e^{-rt} \geq -M, \tag{8}$$

and that for all t, θ, k and all controls $u \geq 0$

$$\lim_{T \to \infty} E^{t,\theta,k}[h(Y_T^u)] = 0, \tag{9}$$

where the state of the system, Y_t, is defined by $Y_t = (t, \Theta_t, K_t)$. Then $h = H$, where H is the optimal value function defined in (1), and the optimal policy is

$$u^*(t, \theta, k) = \begin{cases} 0 & \text{if } k > \phi(\theta), \\ \infty & \text{if } k \leq \phi(\theta). \end{cases} \tag{10}$$

Proof:
Note that equation (6) gives the inequality

$$\mathcal{L}^v h(t, \theta, k) \leq -(\pi(\theta, k) - vC(k))e^{-rt}. \tag{11}$$

By Dynkin's formula, see e.g. Øksendal (1989), we have for all $T > t$, and for any control u:

$$E^{t,\theta,k}[h(T, \Theta_T, K_T)] = h(t, \theta, k) + E^{t,\theta,k}\left[\int_t^T \mathcal{L}^u h(s, \Theta_s, K_s)ds\right]$$

$$\leq h(t, \theta, k) - E^{t,\theta,k}\left[\int_t^T (\pi(\Theta_s, K_s) \tag{12}\right.$$

$$\left. - u(s, \Theta_s, K_s)C(K_s))e^{-rs}ds\right].$$

Rearranging this gives the inequality:

$$E^{t,\theta,k}\left[\int_t^T (\pi(\Theta_s, K_s) - u(s, \Theta_s, K_s)C(K_s))e^{-rs}ds\right] \leq h(t, \theta, k) \tag{13}$$
$$- E^{t,\theta,k}[h(T, \Theta_T, K_T)].$$

As $T \to \infty$, using (9), the last term in this inequality vanishes, and hence, h dominates the expected profit for any policy u. In other words, $h \geq H$.

To prove equality we proceed as follows: Choose a control $u_s \geq 0$ and $T < \infty$. Then by Dynkin's formula and (6) we have

$$E^{t,\theta,k}[h(Y_T^u)] = h(t,\theta,k) + E^{t,\theta,k}\left[\int_t^T \mathcal{L}^u h(Y_s^u)\,ds\right]$$

$$= h(t,\theta,k) - E^{t,\theta,k}\left[\int_t^T (\pi(\Theta_s, K_s)\right.$$

$$\left. - C(K_s)u_s)e^{-rs}(1 - \chi(Y_s^u))\,ds\right]$$

$$+ E^{t,\theta,k}\left[\int_t^T \mathcal{L}^u h(Y_s^u)\chi(Y_s^u)\,ds\right],$$

where

$$\chi(y) = \chi(t,\theta,k) = \begin{cases} 1 & \text{if } k < \phi(\theta), \\ 0 & \text{if } k \geq \phi(\theta). \end{cases}$$

Defining $J^u(t,\theta,k)$ by the next equation and letting $T \to \infty$ we get

$$J^u(t,\theta,k) = E^{t,\theta,k}\left[\int_t^\infty (\pi(\Theta_s, K_s) - C(K_s)u_s)e^{-rs}ds\right]$$

$$= h(t,\theta,k) + E^{t,\theta,k}\left[\int_t^\infty \{\mathcal{L}^u h(Y_s^u) + (\pi(\Theta_s, K_s)\right.$$

$$\left. - C(K_s)u_s)e^{-rs}\}\chi(Y_s^u)ds\right] \quad (14)$$

$$\geq h(t,\theta,k) - M \cdot E^{t,\theta,k}\left[\int_t^\infty \chi(Y_s^u)ds\right].$$

We now choose $u = w$ defined by

$$w = \begin{cases} m & \text{if } k < \phi(\theta), \\ 0 & \text{if } k \geq \phi(\theta), \end{cases}$$

where m is a large integer. Then observe that if $Y_s^w \in \mathcal{A}$, then K_s increases with speed m. The total amount of time that Y_s^w spends in $\mathcal{A}$ is at most $(\hat{k} - k)/m$, where $\hat{k} = \sup_{\theta \geq 0} \phi(\theta)$. Substituting in (14) with $u = w$ we get

$$J^w(t,\theta,k) \geq h(t,\theta,k) - \frac{M \cdot (\hat{k} - k)}{m} \to h(t,\theta,k) \text{ as } m \to \infty,$$

which shows that $h \leq \sup_u J^u = H$. **QED**

The precise meaning of the singular control u^* as given by (10) is that the corresponding process $Y_t^* = (t, \Theta_t, K_t^*)$ should have no increase in the K_t component (i.e. $u^* = 0$) if (Θ_t, K_t^*) is situated outside $\bar{\mathcal{A}}$, while its K_t component should immediately jump vertically to the boundary $\partial \mathcal{A}$ of $\mathcal{A}$ if (Θ_t, K_t^*) starts inside $\mathcal{A}$. In Kobila (1989) we have shown that (Θ_t, K_t^*) is a Markov process with horizontal movements outside $\mathcal{A}$ and vertical reflection on $\partial \mathcal{A}$.

The generalized Hamilton-Jacobi-Bellman equation (6) can be written as:

$$\sup_{v \geq 0} \left[v \left(\frac{\partial H}{\partial k} - C(k)e^{-rt} \right) + \pi(\theta, k)e^{-rt} \right.$$
$$\left. + \frac{\partial H}{\partial t} + a(\theta)\frac{\partial H}{\partial \theta} + \frac{1}{2}b^2(\theta)\frac{\partial^2 H}{\partial \theta^2} \right] \leq 0 , \tag{15}$$

where equality only is required for $(\theta, k) \notin \mathcal{A}$. Note that the equations (7) and (10) imply that $(\partial H/\partial k) - C(k)e^{-rt} \leq 0$, and that $v = 0$ for $(\partial H/\partial k) - C(k)e^{-rt} < 0$. In both cases the HJB equation (15) can be written:

$$\pi(\theta, k)e^{-rt} + \frac{\partial H}{\partial t} + a(\theta)\frac{\partial H}{\partial \theta} + \frac{1}{2}b^2(\theta)\frac{\partial^2 H}{\partial \theta^2} \leq 0 . \tag{16}$$

Suppose $H(t, \theta, k) = G(\theta, k)e^{-rt}$, then the HJB equation can be written:

$$\pi(\theta, k) - rG + a(\theta)\frac{\partial G}{\partial \theta} + \frac{1}{2}b^2(\theta)\frac{\partial^2 G}{\partial \theta^2} \leq 0 . \tag{17}$$

In $\mathcal{A}$ we have that $\partial G/\partial k = C(k)$. Since marginal profit is increasing in θ, a natural requirement for $\mathcal{A}$ is that it is bounded by the concave curve $k = \phi(\theta)$. This justifies the definition of $\mathcal{A}$ given by (5). Since we know the partial derivative $\partial G/\partial k$ in $\mathcal{A}$, we can express the unknown function G in $\mathcal{A}$ by the value at the boundary by integrating from k to $\phi(\theta)$. Hence,

$$G(\theta, k) = G(\theta, \phi(\theta)) - \int_k^{\phi(\theta)} C(x)dx \qquad \text{for } k < \phi(\theta). \tag{18}$$

Equation (18) has an intuitive economic interpretation. In the area $\mathcal{A}$ we want to increase the capacity to $\phi(\theta)$. This can be done instantaneously, hence the value function at this point is the value in the point we immediately adjust to, less the cost of increasing the capacity to this point.

Using (18) in (17) we need to consider $\partial^2 G/\partial\theta^2$, which is given by

$$
\frac{\partial^2 G}{\partial\theta^2} = \begin{cases} \dfrac{\partial^2 G}{\partial\theta^2}(\theta, k) & \text{for } k > \phi(\theta), \\[2mm] \dfrac{\partial^2 G}{\partial\theta^2}(\theta, \phi(\theta)) + \dfrac{\partial^2 G}{\partial\theta\partial k}(\theta, \phi(\theta))\phi'(\theta) & \text{otherwise.} \end{cases}
$$

In order for $\partial^2 G/\partial\theta^2$ to be continuous at the boundary $k = \phi(\theta)$, we must require that $(\partial^2 G/\partial\theta\partial k)(\theta, \phi(\theta)) = 0$. In this case G must satisfy

$$
-rG + a(\theta)\frac{\partial G}{\partial\theta} + \frac{1}{2}b^2(\theta)\frac{\partial^2 G}{\partial\theta^2} = -\tilde{\pi}(\theta, k)\,, \tag{19}
$$

where

$$
\tilde{\pi}(\theta, k) = \begin{cases} \pi(\theta, k) & \text{for } k > \phi(\theta), \\[2mm] \pi(\theta, \phi(\theta)) - r\int_k^{\phi(\theta)} C(x)dx & \text{otherwise.} \end{cases} \tag{20}
$$

To be able to solve the equation for G we must specify the Itô diffusion Θ_t. Assume from now on that

$$
d\Theta_t = \alpha\Theta_t dt + \beta\Theta_t dB_t\,, \tag{21}
$$

i.e., Θ_t is a geometric Brownian motion. Then the differential equation for G becomes

$$
-rG + \alpha\theta\frac{\partial G}{\partial\theta} + \frac{1}{2}(\beta\theta)^2\frac{\partial^2 G}{\partial\theta^2} = -\tilde{\pi}(\theta, k). \tag{22}
$$

Note that in this differential equation we only have derivatives in θ. k is only introduced at the right hand side, so we have one differential equation for each k. Hence, we only need boundary conditions for $\theta = 0$ and for $\theta = \infty$. Under the assumption of geometric Brownian motion $\Theta_t = 0$ is an absorbing state. A natural requirement is that no capacity expansion should take place when $\theta = 0$. This gives the restriction $\pi'_k(0, k) < rC(k)$. From (22) we have that $G(0, k) = \tilde{\pi}(0, k)/r = \pi(0, k)/r$. The other boundary condition is more complicated. Define

$$
\pi'_k(\infty, k) = \lim_{\theta\to\infty} \pi'_k(\theta, k)\,. \tag{23}
$$

Since π'_k by assumption is increasing in θ this limit exists but may be infinite.

Suppose first that the limit is finite, $\pi'_k(\infty, k) < \infty$. We need to consider the behavior of k for large θ. In Kobila (1989) the existence of an upper bound $k_{\max}$ was secured by a constant opportunity cost. We will here justify the existence of an upper bound $k_{\max}$ for K_t by the following heuristic argument.

Consider an infinitesimal capacity expansion Δk. A minimum requirement for profitable capacity expansion is that the expected marginal increase in profit minus investment cost is positive. Hence, we consider the following expression:

$$
\lim_{\theta \to \infty} E^\theta \int_0^\infty \left[\pi(\Theta_t, k + \Delta k) - \pi(\Theta_t, k) \right] e^{-rt} dt - C(k)\Delta k
$$
$$
\approx \Delta k \lim_{\theta \to \infty} E^\theta \int_0^\infty \pi'_k(\Theta_t, k) e^{-rt} dt - C(k)\Delta k
$$
$$
= \Delta k \left(\frac{\pi'_k(\infty, k)}{r} - C(k) \right) > 0 \,.
$$

The last equality above follows since as $\theta \to \infty$, future revenues are almost certain, and we can disregard the expectation operator and use (23). Hence, we obtain the following inequality as a minimum requirement for profitable capacity expansion,

$$
C(k) < \frac{1}{r}\pi'_k(\infty, k) \,.
$$

Since $C(k)$ is increasing and π'_k is decreasing the function $\eta(k)$ defined by

$$
\eta(k) = C(k) - \frac{1}{r}\pi'_k(\infty, k) \,, \tag{24}
$$

is increasing with exactly one point $k_{\max}$ such that $\eta(k_{\max}) = 0$, see the figure below. We will interpret $k_{\max}$ as the upper bound for K_t as $\theta \to \infty$.

It is not profitable to expand capacity beyond $k_{\max}$, since the cost of this expansion is higher than the upper bound on the expected present value of net future income from the expansion. But as $\theta \to \infty$ it will be optimal to expand the capacity up to a capacity infinitesimally less than $k_{\max}$, since future revenues in this case are almost certain.

Secondly, consider the case when $\pi'_k(\infty, k) = \infty$. Then the existence of an upper bound $k_{\max}$ cannot be justified by a similar heuristic argument. Nevertheless, we will also in this case assume that K_t is bounded by $K_t \leq k_{\max}$.

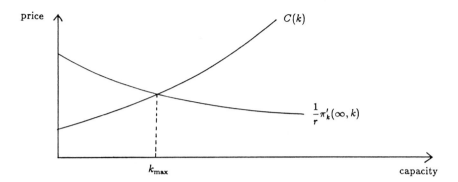

Define $\pi(\infty, k) = \lim_{\theta \to \infty} \pi(\theta, k)$. Assume now that $\pi(\infty, k) < \infty$. A reasonable boundary condition in this case is then

$$
\begin{aligned}
G(\infty, k) &= \lim_{\theta \to \infty} G(\theta, k) \\
&= \lim_{\theta \to \infty} E^\theta \left\{ \int_0^\infty \pi(\Theta_t, k_{\max}) e^{-rt} dt \right\} - \int_k^{k_{\max}} C(x) dx \\
&= \frac{\pi(\infty, k_{\max})}{r} - \int_k^{k_{\max}} C(x) dx \\
&= \frac{\tilde{\pi}(\infty, k)}{r} .
\end{aligned}
\tag{25}
$$

We have here used that when $\theta \to \infty$, $K_t \to k_{\max}$, and future revenues are almost certain so that the expectation operator can be disregarded.

In the case where $\pi(\infty, k) = \infty$, the boundary condition can only be explicitly stated in specific cases. As before we need that K_t is bounded by $K_t \leq k_{\max}$. Consider the following specification of the profit function,

$$
\pi(\theta, k) = \theta \lambda(k) - \xi(k) ,
\tag{26}
$$

where $\lambda'(k) > 0, \lambda''(k) < 0$ and $\xi'(k) > 0$. Here $\xi(k)$ represents unit cost. As explained above, for large θ the optimal policy is to invest until

$k \approx k_{\max}$, and the value function becomes

$$
G(\theta, k) \approx \overline{\lambda} \cdot E^\theta \left\{ \int_0^\infty \Theta_t e^{-rt} dt \right\}
$$
$$
- \int_0^\infty \xi(k) e^{-rt} dt - \int_k^{k_{\max}} C(x)\, dx \qquad (27)
$$
$$
= \overline{\lambda} \frac{\theta}{r - \alpha} - N \,,
$$

where N is independent of θ, and we use the notation $\overline{\lambda} = \lambda(k_{\max})$. We have here used that the expected value, $E^\theta[\Theta_t]$, when Θ_t is given by (21), is $\theta e^{\alpha t}$. Dividing by θ and taking the limit for $\theta \to \infty$, we get

$$
\lim_{\theta \to \infty} \frac{G(\theta, k)}{\theta} = \frac{\overline{\lambda}}{r - \alpha} \quad \text{for } k \le k_{\max} \,. \qquad (28)
$$

The solution to (22) with boundary conditions (25) and (28) is given by the following lemma.

Lemma 1. Assume that $f : [0, \infty) \to \mathbf{R}$ is a continuous function such that either

(i) $\lim_{\theta \to \infty} f(\theta) = f(\infty)$ exists or
(ii) $\lim_{\theta \to \infty} f(\theta)/\theta = -\overline{\lambda}$ exists.

Then there exists a unique solution to the differential equation:

$$
-rg + \alpha\theta g'(\theta) + \frac{1}{2}(\beta\theta)^2 g''(\theta) = f(\theta) \,, \qquad (29)
$$

such that

$$
g(0) = -\frac{f(0)}{r} \,,
$$

and

$$
\lim_{\theta \to \infty} g(\theta) = -\frac{f(\infty)}{r} \qquad \text{in case (i)} \,,
$$

or

$$
\lim_{\theta \to \infty} \frac{g(\theta)}{\theta} = \frac{\overline{\lambda}}{r - \alpha} \qquad \text{in case (ii).}
$$

The solution is:

$$
g(\theta) = \frac{2}{(\gamma_1 - \gamma_2)\beta^2} \left[\theta^{\gamma_1} \int_\infty^\theta \frac{f(s)}{s^{\gamma_1 + 1}} ds - \theta^{\gamma_2} \int_0^\theta \frac{f(s)}{s^{\gamma_2 + 1}} ds \right] \,, \qquad (30)
$$

where $\gamma_1 > 0 > \gamma_2$ are the roots of the characteristic equation

$$-r + \left(\alpha - \frac{1}{2}\beta^2\right)\gamma + \frac{1}{2}(\beta\gamma)^2 = 0. \tag{31}$$

Proof:
That $g(\theta)$ is a solution to (29) is easily established by inserting the derivatives into (29) and using the equality (31). To check the limits when $\theta \to \infty$, we use L'Hôpital's rule. In case (i) we get

$$\lim_{\theta\to\infty}\theta^{\gamma_1}\int_\infty^\theta \frac{f(s)}{s^{\gamma_1+1}}ds = \lim_{\theta\to\infty}\frac{\int_\infty^\theta \frac{f(s)}{s^{\gamma_1+1}}ds}{\theta^{-\gamma_1}}$$

$$= \lim_{\theta\to\infty}\frac{\frac{f(\theta)}{\theta^{\gamma_1+1}}}{-\gamma_1\theta^{-\gamma_1-1}} = -\frac{f(\infty)}{\gamma_1},$$

$$\lim_{\theta\to\infty}\theta^{\gamma_2}\int_0^\theta \frac{f(s)}{s^{\gamma_2+1}}ds = \lim_{\theta\to\infty}\frac{\int_0^\theta \frac{f(s)}{s^{\gamma_2+1}}ds}{\theta^{-\gamma_2}}$$

$$= \lim_{\theta\to\infty}\frac{\frac{f(\theta)}{\theta^{\gamma_2+1}}}{-\gamma_2\theta^{-\gamma_2-1}} = -\frac{f(\infty)}{\gamma_2}.$$

Hence,

$$\lim_{\theta\to\infty}g(\theta) = \frac{2}{(\gamma_1-\gamma_2)\beta^2}\left(-\frac{f(\infty)}{\gamma_1}+\frac{f(\infty)}{\gamma_2}\right)$$

$$= \frac{2f(\infty)}{\gamma_1\gamma_2\beta^2} = -\frac{f(\infty)}{r}, \tag{32}$$

since $\gamma_1\gamma_2 = -2r/\beta^2$. The condition $g(0) = -f(0)/r$ is checked in the same way. Similarly, in case (ii) we find

$$\lim_{\theta\to\infty}\frac{1}{\theta}\theta^{\gamma_1}\int_\infty^\theta \frac{f(s)}{s^{\gamma_1+1}}ds = \lim_{\theta\to\infty}\frac{\int_\infty^\theta \frac{f(s)}{s^{\gamma_1+1}}ds}{\theta^{-(\gamma_1-1)}}$$

$$= \lim_{\theta\to\infty}\frac{\frac{f(\theta)}{\theta^{\gamma_1+1}}}{-(\gamma_1-1)\theta^{-\gamma_1}}$$

$$= \lim_{\theta\to\infty}\frac{1}{1-\gamma_1}\frac{f(\theta)}{\theta} = \frac{\bar{\lambda}}{\gamma_1-1},$$

$$\lim_{\theta\to\infty}\frac{1}{\theta}\theta^{\gamma_2}\int_0^\theta\frac{f(s)}{s^{\gamma_2+1}}ds = \lim_{\theta\to\infty}\frac{\int_0^\theta\frac{f(s)}{s^{\gamma_2+1}}ds}{\theta^{-(\gamma_2-1)}}$$

$$=\lim_{\theta\to\infty}\frac{\frac{f(\theta)}{\theta^{\gamma_2+1}}}{-(\gamma_2-1)\theta^{-\gamma_2}}$$

$$=\frac{1}{1-\gamma_2}\frac{f(\theta)}{\theta}=\frac{\bar\lambda}{\gamma_2-1}.$$

Hence,

$$\lim_{\theta\to\infty}\frac{g(\theta)}{\theta}=\frac{2}{(\gamma_1-\gamma_2)\beta^2}\left(\frac{\bar\lambda}{\gamma_1-1}-\frac{\bar\lambda}{\gamma_2-1}\right)$$

$$=-\frac{2(\gamma_2-\gamma_1)\bar\lambda}{(\gamma_1-\gamma_2)\beta^2(r-\alpha)\frac{2}{\beta^2}}=\frac{\bar\lambda}{r-\alpha}. \tag{33}$$

Here we have used that

$$(\gamma_1-1)(\gamma_2-1)=-\frac{2}{\beta^2}(r-\alpha),$$

which follows from

$$\gamma_1\gamma_2=-\frac{2r}{\beta^2}\quad\text{and}\quad\gamma_1+\gamma_2=1-\frac{2\alpha}{\beta^2}.$$

QED

In order to completely characterize the solution, we need to determine the boundary $k=\phi(\theta)$. It is more convenient to consider the inverse function. Let $\psi(k)$ denote the inverse of $\phi(\theta)$, and note that for $\theta=\psi(k)$ we have

$$C(k)=G'_k(\psi(k),k). \tag{34}$$

Assume that $f(\theta)=-\tilde\pi(\theta,k)$ satisfies (i) or (ii) of Lemma 1, for each k. Then we can apply Lemma 1 to (22) and insert the solution given by (30). Using (20), this gives

$$C(k)=G'_k(\psi(k),k)$$

$$=\frac{2}{(\gamma_1-\gamma_2)\beta^2}\left[-\psi(k)^{\gamma_1}\int_\infty^{\psi(k)}\frac{\tilde\pi'_k(s,k)}{s^{\gamma_1+1}}ds\right.$$

$$\left.+\psi(k)^{\gamma_2}\int_0^{\psi(k)}\frac{\tilde\pi'_k(s,k)}{s^{\gamma_2+1}}ds\right]$$

$$= \frac{2}{(\gamma_1 - \gamma_2)\beta^2} \left[\psi(k)^{\gamma_1} \int_{\psi(k)}^{\infty} \frac{\tilde{\pi}'_k(s,k)}{s^{\gamma_1+1}} ds \right.$$

$$\left. + \psi(k)^{\gamma_2} \int_0^{\psi(k)} \frac{\tilde{\pi}'_k(s,k)}{s^{\gamma_2+1}} ds \right]$$

$$= \frac{2}{(\gamma_1 - \gamma_2)\beta^2} \left[\psi(k)^{\gamma_1} \int_{\psi(k)}^{\infty} \frac{rC(k)}{s^{\gamma_1+1}} ds \right.$$

$$\left. + \psi(k)^{\gamma_2} \int_0^{\psi(k)} \frac{\pi'_k(s,k)}{s^{\gamma_2+1}} ds \right] \tag{35}$$

$$= \frac{2}{(\gamma_1 - \gamma_2)\beta^2} \left[\frac{rC(k)}{\gamma_1} + \psi(k)^{\gamma_2} \int_0^{\psi(k)} \frac{\pi'_k(s,k)}{s^{\gamma_2+1}} ds \right],$$

where the last equality follows from

$$\int_{\psi(k)}^{\infty} s^{-\gamma_1-1} ds = \psi(k)^{-\gamma_1}/\gamma_1.$$

In order to simplify this expression, we rewrite (35) and obtain

$$C(k)\left[1 - \frac{2r}{(\gamma_1 - \gamma_2)\beta^2\gamma_1}\right] = \frac{2}{(\gamma_1 - \gamma_2)\beta^2} \psi(k)^{\gamma_2} \int_0^{\psi(k)} \frac{\pi'_k(s,k)}{s^{\gamma_2+1}} ds.$$

Rearranging, we get

$$C(k) = \frac{2\gamma_1}{(\gamma_1 - \gamma_2)\beta^2\gamma_1 - 2r} \psi(k)^{\gamma_2} \int_0^{\psi(k)} \frac{\pi'_k(s,k)}{s^{\gamma_2+1}} ds.$$

Using the equality $\gamma_1\gamma_2 = -2r/\beta^2$ from Lemma 1, we obtain

$$C(k) = \frac{2}{\beta^2\gamma_1} \psi(k)^{\gamma_2} \int_0^{\psi(k)} \frac{\pi'_k(s,k)}{s^{\gamma_2+1}} ds. \tag{36}$$

We state this important equation as a theorem.

Theorem 2. Let Θ_t be a geometric Brownian motion as given by (21). Assume that $\pi(\infty, k) < \infty$ and that K_t is bounded by $K_t \leq k_{\max}$. Moreover, assume that there exists $M < \infty$ such that

$$\tilde{\pi}(\theta, k) - \pi(\theta, k) \leq M, \tag{37}$$

where $\tilde{\pi}(\theta, k)$ is defined in (20). Then the solution to

$$H(t, \theta, k) = \sup_{u_s} E^{t,\theta,k}\left[\int_t^\infty (\pi(\Theta_s, K_s) - C(K_s)u_s)e^{-rs}ds\right] \qquad (38)$$

is given by $H = h$, where

$$h(t, \theta, k) = \frac{2e^{-rt}}{(\gamma_1 - \gamma_2)\beta^2}\left[-\theta^{\gamma_1}\int_\infty^\theta \frac{\tilde{\pi}(s, k)}{s^{\gamma_1+1}}ds\right.$$
$$\left. +\theta^{\gamma_2}\int_0^\theta \frac{\tilde{\pi}(s, k)}{s^{\gamma_2+1}}ds\right]. \qquad (39)$$

The corresponding optimal control is

$$u^*(t, \theta, k) = \begin{cases} 0 & \text{for } \theta < \psi(k), \\ \infty & \text{otherwise,} \end{cases} \qquad (40)$$

where ψ is determined by the equation

$$C(k) = \frac{2}{\beta^2\gamma_1}\psi(k)^{\gamma_2}\int_0^{\psi(k)} \frac{\pi'_k(s, k)}{s^{\gamma_2+1}}ds. \qquad (41)$$

Proof:
As shown in (25) the function $f(\theta) = -\tilde{\pi}(\theta, k)$ satisfies (i) of Lemma 1 for each k. So for each k the function $g(\theta, k)$ defined by

$$h(t, \theta, k) = e^{-rt}g(\theta, k) \qquad (42)$$

solves (22) by Lemma 1. We have shown that (22) is equivalent to (6) and (7) of Theorem 1. So (39) and (40) follow from Theorem 1 and then (41) follows from (36) once we have established that (8) holds.
Now if $k < \phi(\theta)$ we have

$$\mathcal{L}^v h + (\pi - vC)e^{-rt} = \left[-rg + \alpha\theta\frac{\partial g}{\partial \theta} + \frac{1}{2}(\beta\theta)^2\frac{\partial^2 g}{\partial \theta^2}\right.$$
$$\left. +\pi v\left(\frac{\partial g}{\partial k} - C\right)\right]e^{-rt} \qquad (43)$$
$$= \left[-\tilde{\pi}(\theta, k) + \pi(\theta, k)\right]e^{-rt},$$

because $\partial g/\partial k = C(k)$, by (42) and the definition of h. Hence (8) follows from (37) and the proof is complete. **QED**

Despite the general form of Theorem 2 it gives an investment strategy with a simple intuitive interpretation. The optimal investment rate is zero as long as the random variable Θ_t is below a critical level $\psi(K_t)$. The critical level depends on $\pi'_k(\theta, k)$, which represents the expected future income potential from a marginal capacity expansion. When the random variable Θ_t is sufficiently high, capacity is increased according to the infinite investment rate.

When $\pi(\theta, k)$ is a linear function in θ, (37) of Theorem 2 is not satisfied, and we need to impose a different restriction. We state the result for the linear case in a separate theorem.

Theorem 3. Let Θ_t be a geometric Brownian motion as given by (21). Suppose that the function $\pi(\theta, k)$,

$$\pi(\theta, k) = \theta\lambda(k) - \xi(k), \tag{44}$$

is linear in θ, with $\lambda'(k) > 0, \lambda''(k) < 0$ and $\xi'(k) > 0$. Suppose that the stochastic process is restricted by $K_t \leq k_{\max}$. Then the solution to

$$H(t, \theta, k) = \sup_{u_s} E^{t,\theta,k}\left[\int_t^\infty (\pi(\Theta_s, K_s) - C(K_s)u_s)e^{-rs}ds\right]$$

is given by $H = h$, where

$$h(t, \theta, k) = \frac{2e^{-rt}}{(\gamma_1 - \gamma_2)\beta^2}\left[-\theta^{\gamma_1}\int_\infty^\theta \frac{\tilde{\pi}(s,k)}{s^{\gamma_1+1}}ds \right. $$
$$\left. +\theta^{\gamma_2}\int_0^\theta \frac{\tilde{\pi}(s,k)}{s^{\gamma_2+1}}ds\right]. \tag{45}$$

The corresponding optimal control is

$$u^*(t, \theta, k) = \begin{cases} 0 & \text{for } \theta < \psi(k), \\ \infty & \text{for } \theta \geq \psi(k), \end{cases} \tag{46}$$

where ψ is determined by the equation

$$C(k) = \frac{2}{\beta^2\gamma_1}\psi(k)^{\gamma_2}\int_0^{\psi(k)} \frac{\pi'_k(s,k)}{s^{\gamma_2+1}}ds. \tag{47}$$

Proof:
The proof is parallel to that of Theorem 2, but using case (ii) of Lemma 1. Furthermore, we cannot invoke the results of Theorem 1 directly, since condition (37) is not satisfied. By inspection of the proof of Theorem 1, however, we see that it suffices to show that for controls

$$u = \begin{cases} m & \text{if } k < \phi(\theta), \\ 0 & \text{if } k \geq \phi(\theta), \end{cases} \tag{48}$$

we have

$$\limsup_{m \to \infty} E^{t,\theta,k} \left[\int_t^\infty [\mathcal{L}^u h(Y_s^u) + (\pi(\Theta_s, K_s) \right.$$

$$\left. - u_s C(K_s)) e^{-rs}] \chi(Y_s^u) ds \right] \geq 0. \tag{49}$$

We use (43) to reformulate this in terms of π and $\tilde{\pi}$,

$$\liminf_{m \to \infty} E^{t,\theta,k} \left[\int_t^\infty [\tilde{\pi}(\Theta_s, K_s) - \pi(\Theta_s, K_s)] \chi(Y_s) e^{-rs} ds \right] \leq 0. \tag{50}$$

Note that

$$\tilde{\pi}(\theta, k) - \pi(\theta, k) \leq \theta(\lambda(\hat{k}) - \lambda(k)). \tag{51}$$

Thus it suffices to prove that:

$$\liminf_{m \to \infty} E^{t,\theta,k} \left[\int_t^\infty \Theta_s \chi(Y_s) e^{-rs} ds \right] \leq 0. \tag{52}$$

Using Hölder's inequality and changing the order of integration we get

$$E^{t,\theta,k} \left[\int_t^\infty \Theta_s \chi(Y_s) e^{-rs} ds \right]$$

$$\leq \left[\int_t^\infty E^{t,\theta,k} [(\Theta_s e^{-rs})^p] ds \right]^{(1/p)} E^{t,\theta,k} \left[\int_t^\infty (\chi(Y_s))^q ds \right]^{(1/q)}, \tag{53}$$

with $p > 1$ and $1/p + 1/q = 1$. Using Itô's lemma we see that $(\Theta_s)^p$ is a geometric Brownian motion with drift $p\alpha + (1/2)p(p-1)\beta^2$. Hence,

$$E^{t,\theta,k} [\Theta_s e^{-rs}]^p = \theta e^{-\nu s}, \tag{54}$$

with $\nu = p(r - \alpha) - (1/2)p(p-1)\beta^2$. Since $\nu > 0$ for p sufficiently close to 1, the first term is finite. For the second term we have:

$$E^{t,\theta,k} \left[\int_t^\infty (\chi(Y_s))^q ds \right]^{(1/q)} \leq \left[\frac{\hat{k} - k}{m} \right]^{(1/q)} \to 0. \tag{55}$$

QED

3. A discussion in relation to investment theory

Standard neoclassical investment theory analyzes the investment decision of a profit-maximizing firm with a concave production function. A general conclusion under full certainty, if demand is growing over time, is that capacity should be expanded until the cost of capacity expansion equals the marginal increase in profit. This conclusion can also be extended to the case of uncertain future profit, given that investments are reversible. A corresponding decision rule under uncertainty and irreversibility can be derived from Theorem 2 of the previous section. The concavity of the profit function is captured by Theorem 2. Integrating (41) by parts gives

$$rC(k) = \pi'_k(\psi(k), k) - \int_0^{\psi(k)} \left(\frac{\psi(k)}{s}\right)^{\gamma_2} \pi''_{\theta k}(s, k)ds. \tag{56}$$

Under full certainty and reversible investments we have the familiar optimality condition that $rC(k) = \pi'_k$. The second term in (56) expresses the effect of uncertainty and irreversibility, which contributes towards a lower optimal capacity. Under full certainty (i.e. $\beta = 0$) we have that $\gamma_2 = -\infty$, hence, the integral in (56) is zero, and the condition $rC(k) = \pi'_k$ is obtained as the limiting case.

Note that in the special case of $\pi''_{\theta k} = 0$, the integral in (56) is zero, and the optimal investment strategy under uncertainty and irreversibility coincides with the certainty case.

In the case of a profit function which is linear in θ we obtain an optimality condition which is similar to (56). Integrating (47) using (44) gives

$$\begin{aligned} rC(k) &= \pi'_k(\psi(k), k) - \lambda'(k)\frac{1}{1-\gamma_2}\psi(k) \\ &= \lambda'(k)\psi(k)\frac{-\gamma_2}{1-\gamma_2} - \xi'(k). \end{aligned} \tag{57}$$

Compared to the case of full certainty the marginal revenue is adjusted by the factor $-\gamma_2/(1-\gamma_2) < 1$.

Rewriting (57) we find the following expression for $\psi(k)$,

$$\psi(k) = (rC(k) + \xi'(k))\frac{1}{\lambda'(k)}\frac{1-\gamma_2}{-\gamma_2}. \tag{58}$$

The interpretation of (58) is that the reservation price $\psi(k)$ equals the sum of the rental cost $rC(k)$ and marginal variable cost $\xi'(k)$, weighted by the marginal productivity $\lambda'(k)$, and weighted by the factor $(1-\gamma_2)/(-\gamma_2) > 1$, which represents risk-adjustment and the irreversibility premium. This is a certainty equivalent result in the sense that the optimality condition under uncertainty and irreversibility is of the same form as the optimality condition under full certainty. The difference consists of the factor $(1 - \gamma_2)/(-\gamma_2) > 1$, which leads to a smaller optimal capacity under uncertainty and irreversibility. In the limiting case of $\beta = 0$ (full certainty), $\gamma_2 = -\infty$ and the risk-adjustment vanishes.

Finally, we will discuss the relationship between the solution in Theorem 3 and the investment model in Pindyck (1988), and show that his solution is a special case of our model. Pindyck (1988) analyzes the problem of optimal capacity choice and capacity expansion under uncertainty in future demand and irreversibility of investment. He explicitly states the problem of evaluating a marginal unit of capacity as an option value problem and establishes the link to financial option value techniques.

Consider a firm facing a linear demand function where the product price P_t depends on a random variable Θ_t and the capacity K_t,

$$P_t = \Theta_t - \mu K_t, \tag{59}$$

where $\mu > 0$. The investment cost function is assumed quadratic,

$$C(K_t) = c_1 K_t + \frac{1}{2} c_2 K_t^2. \tag{60}$$

For a given (θ, k) the profit function becomes

$$\begin{aligned}
\pi(\theta, k) &= pk - C(k) \\
&= \theta k - \mu k^2 - c_1 k - \frac{1}{2} c_2 k^2.
\end{aligned} \tag{61}$$

In order to use Theorem 3, we have to assume a capacity constraint $K_t \le k_{\max}$. We then get

$$\pi'_k(\theta, k) = \max[0, \theta - \zeta(k)], \tag{62}$$

where $\zeta(k) = (2\mu + c_2)k + c_1$. Since (62) is linear for large θ, we can show that we can apply Theorem 3 to this specification of marginal profit. Due to the maximum operator in (62), the following expression for the critical value $C(k)$ will differ slightly from what we found above in (57). Using

(47) with (62) and the relations between the roots of the characteristic equation, we find

$$C(k) = \frac{2}{\gamma_1 \beta^2} \psi(k)^{\gamma_2} \int_{\zeta(k)}^{\psi(k)} \frac{s - \zeta(k)}{s^{\gamma_2+1}} ds$$

$$= \frac{2}{\gamma_1 \beta^2} \left[\frac{\psi(k)}{1 - \gamma_2} - \psi(k)^{\gamma_2} \frac{\zeta(k)^{1-\gamma_2}}{\gamma_2(1 - \gamma_2)} + \frac{\zeta(k)}{\gamma_2} \right] \qquad (63)$$

$$= -\psi(k)\frac{\gamma_2}{r(1 - \gamma_2)} + \psi(k)^{\gamma_2}\zeta(k)^{1-\gamma_2} \frac{1}{r(1 - \gamma_2)} - \frac{\zeta(k)}{r}.$$

We will now show that the critical value for investment, $C(k)$, as given by (63), coincides with the critical value found by Pindyck (1988). In our notation Pindyck's solution (his eq. (11)) is given by

$$C(k) = \psi(k)\frac{(\gamma_1 - 1)}{\gamma_1(r - \alpha)} + \psi(k)^{\gamma_2}\zeta(k)^{1-\gamma_2} \frac{(r - \gamma_1\alpha)}{\gamma_1 r(r - \alpha)} - \frac{\zeta(k)}{r}. \qquad (64)$$

Note that we use a risk-free discount rate whereas Pindyck uses a risk-adjusted discount rate derived from the Capital Asset Pricing Model. To show that (64) equals (63) note that the coefficients of $\zeta(k)$ coincide, and check the coefficients of $\psi(k)$ and $\psi(k)^{\gamma_2}\zeta(k)^{1-\gamma_2}$, again using the relations between the roots of the characteristic equation, cf. the proof of Lemma 1. In particular, note that $r(\gamma_1 - 1)(\gamma_2 - 1) = \gamma_1\gamma_2(r - \alpha)$. By comparing (63) and (64) term by term we find:

$$\text{For } \psi(k): \quad -\frac{\gamma_2}{r(1 - \gamma_2)} = \frac{\gamma_2\gamma_1(\gamma_1 - 1)}{r\gamma_1(\gamma_1 - 1)(\gamma_2 - 1)} = \frac{\gamma_1 - 1}{\gamma_1(r - \alpha)}.$$

$$\text{For } \psi(k)^{\gamma_2}\zeta(k)^{1-\gamma_2}: \quad \frac{1}{r(1 - \gamma_2)} = \frac{1 - \gamma_1}{r(1 - \gamma_1)(1 - \gamma_2)}$$

$$= \frac{\gamma_1\left(\gamma_2 + \dfrac{2\alpha}{\beta^2}\right)}{\gamma_1^2\gamma_2(r - \alpha)} = \frac{r - \gamma_1\alpha}{\gamma_1 r(r - \alpha)}.$$

Hence, (63) and (64) coincide, and we have shown that Pindyck's solution corresponds to our Theorem 3.

We have thus shown that a stochastic dynamic programming approach yields the same solution as an explicit option value problem. The intuition behind this result is quite straightforward. Since the control problem is linear in u_t, the rate of investment, the solution becomes singular, i.e. the investment rate is either zero or infinity. The interpretation is that the stochastic control problem degenerates into what is formally an optimal stopping problem, or equivalently, an option value problem.

References

Abel A.B. (1983): "Optimal Investment under Uncertainty", *American Economic Review* 73: 228–233.

Arrow K.J. (1968): "Optimal Capital Policy with Irreversible Investment", in J.N. Wolfe (editor), *Value, Capital and Growth, Papers in Honour of Sir John Hicks*, Edinburgh University Press.

Brennan M.J., and E.S. Schwartz (1985): "Evaluating Natural Resource Investments", *Journal of Business* 58: 135–157.

Kobila T.Ø. (1989): "An Application of Reflected Diffusions to the Problem of Choosing between Hydro and Thermal Power Generation", Preprint No. 5, Department of Mathematics, University of Oslo.

McDonald R., and D. Siegel (1986): "The Value of Waiting to Invest", *Quarterly Journal of Economics* 101: 707–727.

Nickell S.J. (1978): *The Investment Decisions of Firms*, Cambridge University Press.

Øksendal B. (1989): *Stochastic Differential Equations. An Introduction with Applications*, second edition, Springer-Verlag.

Pindyck R.S. (1988): "Irreversible Investment, Capacity Choice and the Value of the Firm", *American Economic Review* 78: 969–985.

Stochastic Models and Option Values
D. Lund and B. Øksendal (Editors)
© Elsevier Science Publishers B.V. (North-Holland), 1991

The High Contact Principle as a Sufficiency Condition for Optimal Stopping

Kjell Arne Brekke[1] and Bernt Øksendal[2]

[1]*Research Department*
Central Bureau of Statistics
N-0033 Oslo 1, Norway
[2]*Department of Mathematics, University of Oslo*
N-0316 Oslo 3, Norway

1. Introduction

The "high contact" principle was first introduced by Samuelson (1965). He only gave a heuristic argument for the condition. In McKean's mathematical appendix to Samuelson's paper (McKean (1965)), a rigorous proof for the necessity of the condition was given for the case of a linear reward and a geometrical Brownian motion process, i.e., he proved that any solution of the optimal stopping problem has to satisfy the high contact condition. As there only exists one function satisfying the high contact condition in this situation, Samuelson's proposal is the only possible optimal solution.

In Shiryayev (1978, Theorem 3.17) the condition appears as a necessary condition for one dimensional processes (but not under the name "high contact"). Multidimensional versions of the theorem are given in Grigolionis and Shiryayev (1966), Friedman (1976) and Bensoussan and Lions (1982). However, the assumptions in these theorems seem to be too strong to apply to most economic applications. (See section 2.)

The high contact principle is essentially a first order condition in the optimization of the stopping time (see Merton (1973, footnote 60), and for a rigorous further development of the same idea Øksendal (1990)). To derive the "second order conditions" turns out to be easy, and we will in this paper prove, under weak conditions, that a solution proposal to an optimal stopping problem satisfying the high contact principle, is in fact an optimal solution to the problem. In this case we do not have to prove

We thank Trond Olsen and Henrik H. Martens for comments to the first version of this paper.

that there is only one solution satisfying the high contact principle, and the existence of the optimal solution is a part of the conclusion.

Some results from stochastic analysis will be used without reference, these results can be found in Øksendal (1989).

2. The problem

Let

$$dX_t = b(X_t)dt + \sigma(X_t)dB_t \tag{1}$$

be an n-dimensional Itô diffusion, where $b : \mathbf{R}^n \mapsto \mathbf{R}^n$ and $\sigma : \mathbf{R}^n \mapsto \mathbf{R}^{n \times m}$ are Lipschitz continuous functions with at most linear growth. Let g be a (real) bounded continuous function on $\mathbf{R}^n$. The optimal stopping problem is the problem of finding:

$$g^*(x) = \sup_\tau E^x[g(X_\tau)], \tag{2}$$

the sup being taken over all $\mathcal{F}_t$-stopping times τ, where $\mathcal{F}_t$ is the σ-algebra generated by B_s, $s \leq t$. Here E^x denotes the expectation w.r.t. the law P^x of $\{X_t\}$ given $X_0 = x$. The *optimal stopping time* corresponding to g^* is denoted τ^*.

If we know g^* it is easy to find τ^*: It is obviously optimal to stop if $g(X_t) \geq g^*(X_t)$ since we then achieve the optimal benefit, while if $g(X_t) < g^*(X_t)$ it is not optimal to stop, since this would give less than the optimum. The set

$$D = \{x : g^*(x) > g(x)\}$$

is called the continuation region. Obviously, for $x \in D$,

$$g^*(x) = E^x\{g(X_{\tau_D})\},$$

where $\tau_D = \inf\{s > t : X_s \notin D\}$.

If the continuation region is known, then the problem of finding g^* can be transformed into a Dirichlet problem. Define the operator

$$L = \sum_i b_i(x)\frac{\partial}{\partial x_i} + \frac{1}{2}\sum_{i,j} a_{i,j}(x)\frac{\partial^2}{\partial x_i \partial x_j}, \tag{3}$$

where $a = \sigma\sigma^T$, and where b and σ are as in (1). Then it is known that g^* solves the Dirichlet problem:

$$(Lg^*)(x) = 0 \qquad \text{for } x \in D\,,$$
$$\lim_{x \to y} g^*(x) = g(y) \quad \text{for all regular } y \in \partial D. \tag{4}$$

($y \in \partial D$ is called regular if $\tau_D = 0$ a.s. P^y.)

But D is unknown, so this is a free boundary problem. Therefore an additional boundary condition is needed to identify the boundary. This is why the "high contact" principle is important. The principle states that

$$\nabla g^* = \nabla g \qquad \text{on } \partial D\,, \tag{5}$$

which gives us the extra boundary condition.

In a footnote Merton (1969) derives the high contact principle as a first order condition. Suppose X_t is one dimensional. Let $D_c = \{x : x < c\}$, $\tau_c = \tau_{D_c}$, and

$$f(x, c) = E^x[g(X_{\tau_c})]\,. \tag{6}$$

Note that if c is regular for D_c then $f(c, c) = g(c)$, and hence if f'_c exists, we have

$$f'_x(c, c) + f'_c(c, c) = g'(c)\,. \tag{7}$$

Suppose

$$g^*(x) = f(x, c^*) = \max_c f(x, c),$$

then the high contact condition $f'_x(c^*, c^*) = g'(c^*)$ is a direct consequence of the first order condition $f'_c(x, c^*) = 0$.

To make this argument rigorous it is necessary to verify that:

(i) D is of the form D_c, and c is a regular point for D_c.
(ii) $c \mapsto f(x, c) \in C^1$.

And if we want to use this high contact principle to prove that a proposed solution is optimal we must show that:
(iii) The candidate satisfying high contact is unique.
(iv) There exists an optimal solution.

In the multidimensional case, which is the most relevant one for economic applications, it is necessary to have smoothness assumptions on the *optimal* continuation regions as well. In Øksendal (1990) a property analogous to (ii) is verified under the assumption that L is elliptic. Since the argument only proves the *necessity* of high contact, we must also verify (iii) and (iv).

The purpose of this paper is to show that — under certain conditions — the high contact property is also *sufficient* for the solution of the optimal stopping problem. More precisely, we show that if there exists an open set $D \subset \mathbf{R}^n$ with C^1-boundary and a function h on D such that

$$h \geq g \qquad \text{on } D, \tag{8}$$

$$Lg \leq 0 \qquad \text{outside the closure } \overline{D}, \tag{9}$$

("the second order condition") and such that (D, h) solves the free boundary problem

$$Lh = 0 \qquad \text{on } D, \tag{10}$$

$$h = g \qquad \text{on } \partial D, \tag{11}$$

$$\nabla h = \nabla g \quad \text{on } \partial D, \tag{12}$$

then in fact $h = g^*$ on D.

To achieve this result the basic idea is the following. Extend h to $\mathbf{R}^n$ by setting $h = g$ outside D. We know that g^* is the least superharmonic majorant of g. We also know that $h \leq g^*$ since h is what we get from using $\tau = \tau_D$. It only remains to show that h is X_t-superharmonic, i.e:

$$h(x) \geq E^x[h(X_\tau)] \tag{13}$$

for all stopping times τ, and all x. If $h \in C^2$, then this is equivalent to $Lh \leq 0$. This follows from Dynkin's formula:

$$h(x) = E^x \left[h(X_\tau) - \int_0^\tau Lh(X_t)dt \right]. \tag{14}$$

By construction $Lh = 0$ in D, and by assumption $Lh \leq 0$ outside the closure $\overline{D}$. Unfortunately we generally only have $h \in C^1(\partial D)$ (this is the high contact principle). If we can approximate h with $\hat{h} \in C^2$ such that $|h - \hat{h}| < \epsilon$ and $L\hat{h} \leq \epsilon$, for an arbitrary small ϵ, this will do. In Brekke (1989, Appendix C) this idea is used to prove the sufficiency of the high contact in an essentially one-dimensional case.

To extend this result to the multidimensional case, it turns out to be more convenient to generalize the Dynkin formula, using a Green function. This is done in Section 4, but first we will consider the relevance of the theorem in economic applications.

3. High contact in economics

As we have pointed out in the introduction, the existing high contact theorems are either one-dimensional or make very strong assumptions. Let us consider a typical problem in the economics of exhaustible resources, the problems will be similar in other economic applications. Let

$$dP_t = \alpha P_t dt + \beta P_t dB_t \tag{15}$$

be the price of the resource, where α and β are constants. Consider the stopping problem

$$\gamma^*(t,p) = \sup_\tau E^{t,p} \int_t^\tau f(P_s)e^{-rs}ds + g(P_\tau)e^{-r\tau}, \tag{16}$$

where f and g are given bounded continuous functions. If the problem is to find the optimal time to stop production, then f is the profit, and g is the abandonment cost. When the problem is to find the optimal time to start a project, $g(P)$ is the net present value of the field started at price P, and $f = 0$. Note that (16) is not of the form (2), because time and the integral of f until τ are introduced in the reward function. The problem can, however, be brought into the form of (2) at the cost of increasing the dimension of the process. Let $Y_t = (t, P_t, \Theta_t)$ where

$$\Theta_t = \theta + \int_0^t f(P_s)e^{-rs}ds \tag{17}$$

is the "profit" earned until t, then

$$g^*(t,p) + \theta = \sup_\tau E^{t,p,\theta}(g(\tau, P_\tau) + \Theta_\tau). \tag{18}$$

A high-contact principle applying to this problem must allow for processes of dimension three (or two if $f = 0$). In other words, none of the one-dimensional theorems apply to economic problems where discounting is relevant.

Multi-dimensional results derived from regularity results for variational inequalities make assumptions that exclude the geometric Brownian motion (15). A theorem in Friedman (1976) requires e.g. that L is uniformly elliptic and that $a(x) = \sigma\sigma^T$ is bounded, which excludes the geometric Brownian motion where $a(p) = \beta^2 p^2$. Furthermore he assumes that

$$b_i(x) = \sum_{j=1}^n \frac{\partial a_{ij}(x)}{\partial x_j},$$

which in the case of a geometric Brownian motion means $\alpha = 2\beta^2$. These assumptions are too strong for most economic applications.

Another multi-dimensional result is given in Øksendal (1989), who proves the necessity of the high contact principle under the assumption that L is locally elliptic, but he has to make smoothness assumptions on the form of the boundary as well.

Grigolionis and Shiryayev (1966) prove both a necessity and a sufficiency result for high contact in the multi-dimensioned case. Their sufficiency theorem (Theorem 9, p. 555) has weaker assumptions on the stochastic process representing the system, but stronger conditions on the reward function g and the candidate h for g^* than in our paper. It appears to us that it may be difficult to verify these conditions in the economic situations we have in mind.

The aim of this paper is to etablish a sufficiency theorem for high contact which is easy to use and which covers most economic applications. As an illustration we apply it in Section 6 to the problem of starting and stopping of resource extraction.

4. A generalized Dynkin formula

This section uses some advanced mathematical results and methods, and hence it may be difficult to read. Lemma 1 and the following remark contain the results that are needed in the later sections. The reader who is not interested in the details, may skip the remaining part of the section.

Let X_t be as defined in (1) and L the corresponding operator (3). Then X_t has a generator which coincides with L on the smooth functions. Let $L_{\text{loc}}^q(dx)$, $q \geq 1$ be the set of all functions f such that $|f|^q$ is locally integrable with respect to the Lebesgue measure dx, and $C_b^2(A)$ is the set of functions with bounded continuous derivatives of second order in the set A.

If $V \subset \mathbf{R}^n$ is a bounded domain such that $E^\xi[\tau_V] < \infty$ for all ξ then we can define the *Green measure* of X (with respect to V), $\mathcal{G}(\xi, \cdot)$, by

$$\mathcal{G}(\xi, \phi) = \mathcal{G}_V(\xi, \phi) = E^\xi\left[\int_0^{\tau_V} \phi(X_t)dt\right], \qquad \phi \in C(\overline{V}). \qquad (19)$$

If the measure $\mathcal{G}(\xi, dx)$ has the form

$$\mathcal{G}(\xi, dx) = G(\xi, x)dx \qquad (dx \text{ is the Lebesgue measure}), \qquad (20)$$

then we say that X has a *Green function* $G(\xi, x)$ (in V). Note that in this case we can write

$$E^\xi \left[\int_0^{\tau_V} \phi(X_t) dt \right] = \int_V \phi(x) G(\xi, x) dx \,. \tag{21}$$

A sufficient condition that X has a Green function in any bounded domain V, is that X (or, more precisely, the generator L of X) is *uniformly elliptic* in V, i.e. that there exists $\lambda > 0$ such that

$$z^T a(x) z \geq \lambda |z|^2 \qquad \text{for all } x \in V, z \in \mathbf{R}^n \,. \tag{22}$$

In fact, if X is uniformly elliptic in V and $n \geq 3$ then by a result of Littman, Stampacchia and Weinberger (1963) there exists, for any compact H in V, a constant $C = C(H) < \infty$ such that

$$G^X(\xi, x) \leq C \cdot |\xi - x|^{2-n} \qquad \text{for all } \xi, x \in H. \tag{23}$$

Using polar coordinates in $\mathbf{R}^n$ we see that this implies that

$$\int_H G(\xi, x)^q dx = C_1 \int_0^1 r^{-(n-2)q} r^{n-1} dr = C_1 \int_0^1 r^{n-1-q(n-2)} dr \,, \tag{24}$$

which is finite if $q < 1 + n/2$.

We summarize:
If X is uniformly elliptic in V then X has a Green function $G(\xi, x)$ satisfying

$$G(\xi, x) \in L_{\text{loc}}^q(dx) \qquad \text{for } q < 1 + \frac{1}{n} \,.$$

We are also interested in processes of the form

$$dX_t = \begin{bmatrix} dK_t \\ dY_t \end{bmatrix} = \begin{bmatrix} \rho(K_t) \\ \mu(Y_t) \end{bmatrix} dt + \begin{bmatrix} 0 \\ \nu(Y_t) \end{bmatrix} dB_t \,, \tag{26}$$

with $K_t \in \mathbf{R}^m$ and $Y_t \in \mathbf{R}^d$.

Assume for simplicity that $V = M \times N$ where $M \subset \mathbf{R}^m$, $N \subset \mathbf{R}^d$ and assume that Y is uniformly elliptic in N. Then it is well known that Y has a *transition function* $p_t(\eta, y)$ in N, in the sense that

$$E^\eta[\psi(Y_t) \mathcal{X}_{t < \tau_N}] = \int_N \psi(y) p_t(\eta, y) dy, \tag{27}$$

where dy denotes Lebesgue measure (in $\mathbf{R}^d$).

In fact, by a result of Aronson (1967) there exist constants $C < \infty$, $\alpha > 0$ such that

$$p_t(\eta, y) \le C \cdot t^{-d/2} \exp\left(-\frac{\alpha|\eta - y|^2}{t}\right)$$

for all t, and all $x, y \in N$.

Using the transition function of Y we can describe the Green measure $\mathcal{G}_X(\xi, \cdot)$ for X (given by (26)) as follows:

Suppose $\phi(x) = \phi_1(k)\phi_2(y)$ where $x = (k, y)$. Then

$$\mathcal{G}_X(\xi, \phi) = E^\xi\left[\int_0^\infty \phi_1(K_t)\phi_2(Y_t) \cdot \mathcal{X}_{t < \tau_V}\, dt\right]$$
$$= \int_0^\infty \left(\int_N \phi_1(K_t) \cdot \phi_2(y)p_t(\eta, y)dy\right)\mathcal{X}_{K_t \in M}\, dt, \ \xi = (k_0, \eta) \quad (29)$$

Since a general $\phi(k, y)$ can be approximated by a sum of such products, we conclude that for general $\phi(k, y)$ we have

$$\mathcal{G}_X(\xi, \phi) = \int_0^\infty \left(\int_N \phi(K_t, y)p_t(\eta, y)dy\right)\mathcal{X}_{K_t \in M}\, dt, \quad (30)$$

with $p_t(\eta, y)$ satisfying (28), provided that $X_t = (K_t, Y_t)$ where Y_t is uniformly elliptic.

The conclusions (25) and (30) will be needed in the proof of the next result.

Lemma 1. (Generalized Dynkin formula)

Let U, V be bounded domains with C^1 boundaries in $\mathbf{R}^n$ and let Γ denote the boundary of U. Suppose

$$\phi \in C_b^2(V \setminus \Gamma) \cap C^1(\overline{V}), \quad (31)$$

and that X satisfies one of the following two conditions:

1. X_t is uniformly elliptic in V.
2. $X_t = (K_t, Y_t)$ as in (26), with Y_t uniformly elliptic in N (where $V \subset M \times N$) and for each $k \in M$ the set

$$\Gamma_k = \{y \in N : (k, y) \in \Gamma\} \subset \mathbf{R}^d \quad (32)$$

has zero d-dimensional Lebesgue measure.

Then

$$E^\xi[\phi(X_{\tau_V})] = \phi(\xi) + E^\xi\left[\int_0^{\tau_V} L\phi(X_t)dt\right], \tag{33}$$

where $L\phi(x)$ is the function defined for all $x \in V \setminus \Gamma$ by pointwise differentiation according to (3).

Remark: We will use (33) in the form of (44) (in case 1) or (46) (in case 2). In both cases, the expectation is transformed to an integral with respect to the Lebesgue measure. Thus, since the boundary Γ has Lebesgue measure zero, we do not have to define $L\phi$ on the boundary.

Proof: First assume that condition 1. holds.

Let $D_{jk}\phi$ denote the distributional double derivative of ϕ with respect to x_j and x_k and let $\partial^2\phi/\partial x_j\partial x_k$ denote the pointwise double derivative which is defined everywhere outside Γ and hence almost everywhere (dx). We claim that

$$D_{jk}\phi = \frac{\partial^2\phi}{\partial x_j\partial x_k} \qquad \text{for } 1 \le j,k \le n. \tag{34}$$

To establish (34) we put $V_1 = V \cap U$, $V_2 = V \setminus \overline{U}$ and choose $u \in C_0^\infty(V)$. By integration by parts we have

$$\int_{V_i} \frac{\partial^2\phi}{\partial x_j\partial x_k}u\,dx = \int_{\partial V_i} \frac{\partial\phi}{\partial x_k}n_{ij}u\,ds - \int_{V_i} \frac{\partial\phi}{\partial x_k}\frac{\partial u}{\partial x_j}dx, \tag{35}$$

where n_{ij} is component j of the outer unit normal $\vec{n}_i$ from V_i. Another integration by parts leads to

$$\int_{V_i} \frac{\partial\phi}{\partial x_k}\frac{\partial u}{\partial x_j}dx = \int_{\partial V_i} \phi\frac{\partial u}{\partial x_j}n_{ik}ds - \int_{V_i} \phi\frac{\partial^2 u}{\partial x_j\partial x_k}dx. \tag{36}$$

Combining (35), (36) and adding for $i = 1, 2$ gives:

$$\int_V \frac{\partial^2\phi}{\partial x_j\partial x_k}u\,dx = \int_V \phi\frac{\partial^2 u}{\partial x_j\partial x_k}dx, \tag{37}$$

which proves the claim (34).

For $m = 0, 1, 2, \dots$ and $1 \le p < \infty$ define the spaces

$$W^{m,p} = \left\{u : \begin{array}{l} u \in L^p(V); D^\alpha u \in L^p(V) \quad \text{for all multi-indices} \\ \alpha = (\alpha_1, \dots, \alpha_k) \quad \text{with } |\alpha| = \alpha_1 + \cdots + \alpha_k \le m \end{array}\right\}, \tag{38}$$

equipped with the norm

$$\|u\|_{m,p} = \left(\sum_{|\alpha| \leq m} \|D^\alpha u\|_p^p \right)^{1/p} \tag{39}$$

and let $H^{m,p}$ denote the closure of C^m in this norm. Then a famous result of Meyers & Serrin (see e.g. Adams (1975), Th. 3.16) states that

$$H^{m,p} = W^{m,p} . \tag{40}$$

For all $p < \infty$ we have $\phi \in W^{2,p}$, by (31) and (34). So by (40) there exists a sequence $\{\phi_k\} \in C^2$ such that

$$\|\phi_k - \phi\|_{2,p} \to 0 \quad \text{as } k \to \infty . \tag{41}$$

If $p > n/2$ the Sobolev inequality combined with (41) gives that

$$\phi_k \to \phi \quad \text{uniformly on } V . \tag{42}$$

Since $\phi_k \in C^2$, we know that Dynkin's formula holds for ϕ_k, i.e.

$$E^\xi[\phi_k(X_{\tau_V})] = \phi_k(\xi) + \int_V L\phi_k(x)G(\xi,x)dx . \tag{43}$$

Choose $p > n/2 + 1$, then if $1/p + 1/q = 1$ we have $q < 1 + 2/n$, so for such a value of p we can combine (41), (42) and (43) to conclude that:

$$E^\xi[\phi(X_{\tau_V})] = \lim_{k \to \infty} E^\xi[\phi_k(X_{\tau_V})] = \phi(\xi) + \int_V L\phi(x)G(\xi,x)dx , \tag{44}$$

because by Hölder's inequality

$$\left| \int_V (L\phi_k - L\phi)(x)G(\xi,x)dx \right| \leq \|L\phi_k - L\phi\|_p \|G(\xi,\cdot)\|_q \to 0 \tag{45}$$

as $k \to \infty$ by (25). That proves (33).

Next assume that the second condition holds. We proceed as in the first case up to (43), so that we have, using (30), for each k

$$E^\xi[\phi_k(X_{\tau_V})] = \phi_k(\xi) + \int_0^\infty \left(\int_N L\phi_k(K_t,y)p_t(\eta,y)dy \right) \mathcal{X}_{K_t \in M} dt . \tag{46}$$

Choose $1 < q < \infty$ (to be determined later) and apply Hölder's inequality for each t:

$$\int_W |L\phi_k(K_t, y) - L\phi(K_t, y)| \, p_t(\eta, y) dy$$
$$\leq \left(\int_W |L\phi_k - L\phi|^p dy \right)^{1/p} \left(\int_W p_t^q(\eta, y) dy \right)^{1/q} \quad \text{where } \frac{1}{p} + \frac{1}{q} = 1 \,. \tag{47}$$

By the estimate (28) we get, using the substitution $u = y/\sqrt{t}$:

$$\int_W p_t^q(\eta, y) dy \leq C_1 \cdot t^{-qd/2} \int_{\mathbf{R}^d} \exp(-\alpha q |u|^2) t^{d/2} du \leq C_2 \cdot t^{-(q-1)d/2} \,, \tag{48}$$

and this is locally t-integrable near 0 if $q < 1 + 2/d$. (Note that we used the Hölder inequality for each t, so we need a t-uniform approximation, but this follows from the result of Meyers and Serrin which implies that there exist a constant C_3 such that

$$\|\phi_k - \phi\|_{2,p} < C_3 \|\phi\|_{2,p}$$

for all k and all t, where $\|\phi\|_{2,p}$ means the norm of the function $y \mapsto \phi(K_t, y)$.) Therefore, if we choose $1 < q < 1 + 2/d$ and p such that $1/p + 1/q = 1$ we obtain that

$$E^\xi \left[\int_0^{\tau_V} L\phi_k(X_t) dt \right] \to E^\xi \left[\int_0^{\tau_V} L\phi(X_t) dt \right] \tag{49}$$

as before, thereby completing the proof of Lemma 1. **QED**

5. The sufficiency of high contact

We now apply this to the optimal stopping problem (2):

$$g^*(x) = \sup_\tau E^x[g(X_\tau)] = E^x[g(X_{\tau^*})] \,. \tag{50}$$

Theorem 1. (Sufficiency of high contact for the optimal stopping problem)

Suppose $W \subset \mathbf{R}^n$ is an open set such that $X_t \in W$ for all t if $X_0 \in W$. Let $g \in C^1(W)$. Suppose we can find an open set $D \subset W$ with C^1 boundary such that $\tau_D < \infty$ a.s. Assume furthermore that X_t satisfies either condition 1. or 2. of Lemma 1 for $U = D$ and for every sufficiently small open ball V centered on the boundary $\partial D \cup W$, and suppose we can find a function h on $\overline{D}$ such that $h \in C^1(\overline{D}) \cap C^2(D)$, $h \geq g$ on D, $g \in C^2(W \setminus \overline{D})$ and $Lg \leq 0$ in $W \setminus \overline{D}$, and such that (D, h) solves the free boundary problem:

$$
\begin{array}{llll}
\text{(i)} & Lh(x) & = 0 & \text{for } x \in D, \\
\text{(ii)} & h(x) & = g(x) & \text{for } x \in \partial D, \\
\text{(iii)} & \nabla_x h(x) & = \nabla_x g(x) & \text{for } x \in \partial D \cap W \\
& & & \text{if } X_t \text{ satisfies 1.,} \\
\text{(iii)'} & \nabla_y h(x) & = \nabla_y g(x) & \text{for } x \in \partial D \cap W \\
& & & \text{if } X_t \text{ satisfies 2.,}
\end{array}
\tag{51}
$$

(where $x = (k, y)$ if X_t satisfies 2).

Extend h to all of W by putting $h = g$ outside D. Then h solves the optimal stopping problem (50), i.e.:

$$
h(x) = g^*(x) = \sup_\tau E^x[g(X_\tau)],
\tag{52}
$$

and thus $\tau^* = \tau_D$ is an optimal stopping time.

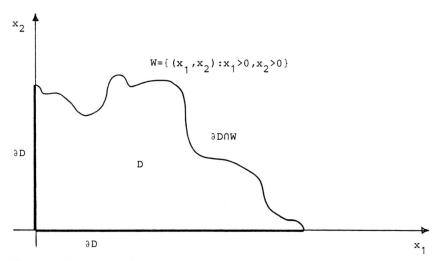

Figure 1: Illustration of boundaries.

Remark: Note that in (51), ∂D is the boundary of the set D regarded as a subset in $\mathbf{R}^n$, not relative to W. Thus there may be parts of ∂D not belonging to W, as illustrated in Figure 1, where W is the positive quadrant of $\mathbf{R}^2$. Note also that (51) (iii) or (iii)' is the high contact condition.

Proof: First we note that by (51) (i), (ii) we have

$$h(x) = E^x[g(X_{\tau_D})] \tag{53}$$

hence $h \leq g^*$. To prove the opposite inequality, it suffices to show that h is X_t-superharmonic since we have assumed that $h \geq g$ and we know that g^* coincides with the least X_t-superharmonic majorant of g. For this it is enough to show that h is *locally* X_t-superharmonic (Dynkin (1965), p. 22). Since $Lh \leq 0$ outside ∂D, h is clearly X_t-superharmonic there. So it remains to show that h is X_t-superharmonic on ∂D:

Fix $\xi \in \partial D$ and let V be a ball centered at ξ. By Lemma 1 we get

$$E^\xi[h(X_\tau)] = h(\xi) + \int_V Lh(x)G(\xi, x)dx \leq h(\xi), \tag{54}$$

where $\tau = \tau_V$. Thus h is locally X_t-superharmonic everywhere and the proof is complete. **QED**

This result also applies to the apparently more general problem

$$\gamma^*(x) = \sup_\tau E^x\left[\int_0^\tau f(X_s)ds + g(X_\tau)\right], \tag{55}$$

where $f(x)$ is a given bounded continuous function. In this case we consider the process Z_t given by:

$$dZ_t = \begin{bmatrix} dX_t \\ d\Theta_t \end{bmatrix} = \begin{bmatrix} b(X_t) \\ f(X_t) \end{bmatrix} dt + \begin{bmatrix} \sigma(X_t) & \vline & 0 \\ 0 & \vline & 1 \end{bmatrix} d\tilde{B}_t, \qquad Z_0 = (x, \theta), \tag{56}$$

where $\tilde{B}_t = (B_1(t), ..., B_{n+1}(t))$ is an $(n+1)$-dimensional Brownian motion. Consider the problem

$$k^*(z) = \sup_\tau E^z[k(Z_\tau)], \tag{57}$$

where $k(z) = k(x, \theta) = g(x) + \theta$.

For $E[\tau] < \infty$ we have $E^0[B_{n+1}(\tau)] = 0$, and since it is enough to take the sup over such stopping times we have

$$k^*(x, \theta) = \sup_\tau E^x \left[\theta + \int_0^\tau f(X_s)ds + B_{n+1}(\tau) + g(X_\tau) \right]$$

$$= \theta + \sup_\tau E^x \left[\int_0^\tau f(X_s)ds + g(X_\tau) \right] \tag{58}$$

$$= \theta + \gamma^*(x).$$

Therefore, if k^* solves the problem (57), then $\gamma^* = k^* - \theta$ solves (55). This gives the following conclusion.

Theorem 2. Suppose $W \subset \mathbf{R}^n$ is an open set such that $X_t \in W$ for all t if $X_0 \in W$. Let $g \in C^1(W)$. Suppose we can find an open set $D \subset W$ with C^1 boundary such that $\tau_D < \infty$ a.s. Assume furthermore that X_t satisfies either condition 1. or 2. of Lemma 1 for $U = D$ and for every sufficiently small open ball V centered on the boundary $\partial D \cup W$, and a function h on $\overline{D}$ such that $h \in C^1(\overline{D}) \cap C^2(D)$, $h \geq g$ on D, $g \in C^2(W \setminus \overline{D})$ and $Lg \leq -f$ in $W \setminus \overline{D}$, and such that (D, h) solves the free boundary problem:

$$
\begin{array}{llll}
\text{(a)} & Lh(x) & = -f(x) & \text{for } x \in D, \\
\text{(b)} & h(x) & = g(x) & \text{for } x \in \partial D, \\
\text{(c)} & \nabla_x h(x) & = \nabla_x g(x) & \text{for } x \in \partial D \cap W \\
& & & \text{if } X_t \text{ satisfies 1.,} \\
\text{(c)}' & \nabla_y h(x) & = \nabla_y g(x) & \text{for } x \in \partial D \cap W \\
& & & \text{if } X_t \text{ satisfies 2.,}
\end{array}
\tag{59}
$$

(where $x = (k, y)$ if X_t satisfies 2).

Suppose furthermore that for all x, we can find $p > 1$, such that:

$$\inf_{\omega \in \Omega, T > 0} \int_0^T f(X_s(\omega))ds > -\infty, \tag{60}$$

where Ω is the measure space on which $\tilde{B}_t$ is defined.

Extend h to all of W by putting $h = g$ outside D. Then h solves the optimal stopping problem (55), i.e.:

$$h(x) = \gamma^*(x) = \sup_\tau E^x \left[\int_0^\tau f(X_s)ds + g(X_\tau) \right], \tag{61}$$

and thus $\tau^* = \tau_D$ is an optimal stopping time.

Proof: Define $H(x, \theta) = h(x) + \theta$, $k(x, \theta) = g(x) + \theta$.

Let

$$\mathcal{L}u(x, \theta) = L_x u + f(x)\frac{\partial u}{\partial \theta} + \frac{1}{2}\frac{\partial^2 u}{\partial \theta^2}, \quad u \in C^\infty \tag{62}$$

denote the generator of the process Z_t.

Then by (a), (b), and (c) we have

$$\mathcal{L}H(x, \theta) = Lh(x) + f(x) = 0 \quad \text{if } x \in D, \tag{63}$$

$$H(x, \theta) = k(x, \theta) \qquad \text{if } x \in \partial D, \tag{64}$$

and

$$\begin{cases} \nabla_x H(x, \theta) = \nabla_x k(x, \theta) \\ \dfrac{\partial H}{\partial \theta}(x, \theta) = \dfrac{\partial k}{\partial \theta}(x, \theta) \end{cases} \quad \text{if } x \in \partial D. \tag{65}$$

By Theorem 1 applied to $h = H$, $g = k$, and the process Z_t, and by (58) we conclude that

$$H(x, \theta) = \sup_\tau E^{x, \theta}[k(Z_\tau)] = \theta + \gamma^*(x), \tag{66}$$

i.e. that $h = \gamma^*$ as claimed.

Note that $k = \theta + g$ is not bounded, hence Theorem 1 does not apply directly. By an inspection of the proof, however, we find that the condition (60) is sufficient. **QED**

Remark: If f is of the form $f(x) = F(y)e^{-rt}$, where $F(y) \geq -M > -\infty$, then

$$\int_0^T f(X_t(\omega))dt \geq -\frac{M}{r} > -\infty, \tag{67}$$

hence (60) is satisfied. This applies to many economic problems.

6. An application: Starting and stopping of resource extraction

The starting and stopping of a mine or a field was studied in a seminal paper by Brennan and Schwartz (1985). For ease of analysis we will simplify their model considerably, and disregard the discussion of taxes and convenience yield. The results in this section is also similar to the entry and exit model of Dixit (1989), where the only difference is that his model includes no resource extraction.

We will formulate the model as two simultaneous stopping problems. The use of the high contact principle in this model illustrates the use of both theorems. Furthermore, we point at an unresolved problem in using optimal stopping theories to solve sequential stopping problems.

Suppose that the price process is a geometric Brownian motion:

$$dP_t = \alpha P_t dt + \beta P_t dB_t \,, \tag{68}$$

where α, β are constants and $\alpha < r$ where r is the interest rate. For convenience we also assume $(1/2)\beta^2 < \alpha$. The stock of remaining reserves in the field is denoted by Q_t. If the field is open, extraction is proportional to remaining reserves. Hence

$$dQ_t = -\mathcal{X}\lambda Q_t dt \,, \tag{69}$$

where

$$\mathcal{X} = \begin{cases} 1 & \text{if the field is open,} \\ 0 & \text{if the field is closed.} \end{cases} \tag{70}$$

When the field is open, a rental cost K is the only operating cost. Thus profit is $\lambda Q_t P_t - K$. It costs C to close the field, and J to open the field.

Let $\tilde{V}$ denote the value of a closed field, and $\tilde{U}$ the value of an open field. Then:

$$\tilde{U}(p,q,t) = \sup_{\tau} \left\{ E^{p,q,t} \left[\int_t^\tau (\lambda Q_s P_s - K) e^{-rs} ds \right. \right.$$
$$\left. \left. + \tilde{V}(P_\tau, Q_\tau, \tau) - C e^{-r\tau} \right] \right\} \,, \tag{71}$$

and

$$\tilde{V}(p,q,t) = \sup_{\tau} \{ E^{p,q,t} [\tilde{U}(P_\tau, Q_\tau, \tau) - J e^{-r\tau}] \} \,. \tag{72}$$

It is reasonable to guess that:

$$\tilde{V}(p,q,t) = V(p,q) e^{-rt} \,, \tag{73}$$
$$\tilde{U}(p,q,t) = U(p,q) e^{-rt} \,. \tag{74}$$

We search for solution proposals v and u that solve the free boundary problem (59). The first condition (a), states that (using the decomposition (73) and (74)):

$$L_V v = 0 \,, \tag{75}$$
$$L_U u = -\lambda qp + K \,, \tag{76}$$

where

$$L_V = -r + \alpha p \frac{\partial}{\partial p} + \frac{1}{2} \beta^2 p^2 \frac{\partial^2}{\partial p^2}, \qquad (77)$$

and

$$L_U = L_V - \lambda q \frac{\partial}{\partial q}. \qquad (78)$$

It is reasonable to guess that the continuation region for the starting problem is of the form $\{(t, p, q) : p < x(q)\}$, and for the stopping problem $\{(t, p, q) : p > y(q)\}$. This gives a boundary at $p = 0$ for v and at $p = \infty$ for u. That $v(0, q) = 0$ is rather obvious.

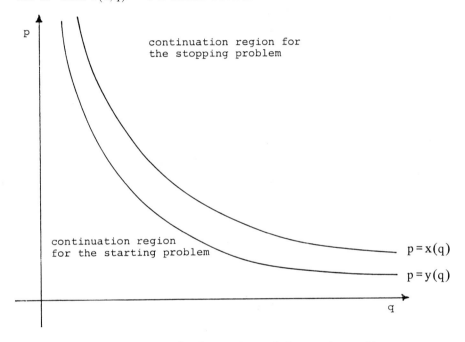

p

continuation region for
the stopping problem

continuation region
for the starting problem

p = x(q)

p = y(q)

q

Figure 2: Continuation region for the starting and the stopping problem.

The boundary condition for u is more complicated. First note that the continuation region D for the stopping problem clearly contains the set

$$D_0 = \{(p, q, t); \ \lambda p q - K > 0\}.$$

So in (71) it suffices to consider τ such that $\tau \geq \tau_0$, where $\tau_0 := \tau_{D_0} := \inf\{t > 0; \ \lambda P_t Q_t \leq K\}$. We claim that, for each q, t and T fixed,

$$P^{p,q,t}[\tau_0 \geq T] \to 1 \quad \text{as} \quad p \to \infty.$$

To prove this put

$$\sigma(a,b) = \inf\{t > 0;\ P_t \notin (a,b)\}, \quad 0 < a < b.$$

Then a short computation (e.g. by solving a Dirichlet problem) shows that

$$P^p[P_{\sigma(a,b)} = a] = \frac{p^\gamma - b^\gamma}{a^\gamma - b^\gamma} \quad \text{where} \quad \gamma = 1 - \frac{2\alpha}{\beta^2} < 0.$$

Hence

$$P^{p,q,t}[\tau_0 \geq T] \geq P^p\left[\sigma\left(\frac{Ke^{rT}}{\lambda q}, \infty\right) \geq T\right]$$

$$\geq P^p\left[\sigma\left(\frac{Ke^{rT}}{\lambda q}, \infty\right) = \infty\right] = 1 - \left(\frac{\lambda pq}{K}e^{-rT}\right)^\gamma \to 1 \quad \text{as} \quad p \to \infty,$$

as claimed.

We conclude that

$$\lim_{p\to\infty} E^{p,q,t}\left[\int_t^\infty (\lambda Q_s P_s - K)e^{-rs}\,ds\right] \leq \lim_{p\to\infty} \tilde{U}(p,q,t)$$

$$\leq \lim_{p\to\infty} E^{p,q,t}\left[\int_t^\infty (\lambda Q_s P_s - K)^+ e^{-rs}\,ds\right],$$

and both the left and the right hand side equals

$$\frac{pq}{r + \lambda - \alpha} - \frac{K}{r}$$

We therefore get the following boundary condition for u:

$$u(p,q) - \left[\frac{pq}{r + \lambda - \alpha} - \frac{K}{r}\right] \to 0 \quad \text{as } p \to \infty. \tag{79}$$

Using these boundary conditions on the equations (75) and (76) we can derive the general form of u and v in the continuation areas. (See Bjerkholt and Brekke (1988).[1])

$$u(p,q) = \frac{pq}{r + \lambda - \alpha} - \frac{K}{r} + c_1 \cdot (pq)^\nu \tag{80}$$

$$v(p,q) = c_2(q) \cdot p^\gamma, \tag{81}$$

where

---

[1] The general form of v was first derived in Olsen and Stensland (1988).

$$\gamma = \frac{-(\alpha - \frac{1}{2}\beta^2) + \sqrt{(\alpha - \frac{1}{2}\beta^2)^2 + 2r\beta^2}}{\beta^2} > 1 \,, \tag{82}$$

$$\nu = \frac{-(\alpha - \lambda - \frac{1}{2}\beta^2) - \sqrt{(\alpha - \lambda - \frac{1}{2}\beta^2)^2 + 2r\beta^2}}{\beta^2} < 0. \tag{83}$$

The boundary conditions at the free boundaries $x(q)$ and $y(q)$ are

$$v(x(q), q) = u(x(q), q) - J \,, \tag{84}$$

$$v_p'(x(q), q) = u_p'(x(q), q) \qquad \text{``high contact''}, \tag{85}$$

$$u(y(q), q) = (y(q), q) - C \,, \tag{86}$$

$$u_p'(y(q), q) = v_p'(y(q), q) \qquad \text{``high contact''}. \tag{87}$$

Inserting the general form of u and v we get

$$\frac{x(q)q}{r + \lambda - \alpha} - \frac{K}{r} + c_1 \cdot (x(q)q)^\nu = c_2(q)(x(q))^\gamma + J \,, \tag{88}$$

$$\frac{x(q)q}{r + \lambda - \alpha} + \nu c_1 \cdot (x(q)q)^\nu = \gamma c_2(q)(x(q))^\gamma \,, \tag{89}$$

$$\frac{y(q)q}{r + \lambda - \alpha} - \frac{K}{r} + c_1 \cdot (y(q)q)^\nu = c_2(q)(y(q))^\gamma - C \,, \tag{90}$$

$$\frac{y(q)q}{r + \lambda - \alpha} + \nu c_1 \cdot (y(q)q)^\nu = \gamma c_2(q)(y(q))^\gamma \,. \tag{91}$$

If we guess that the form of the solution is $x(q) = x/q$, $y(q) = y/q$, and $c_2(q) = k_2 \cdot q^\gamma$, the equation system simplifies to :

$$\frac{x}{r + \lambda - \alpha} + c_1 x^\nu = k_2 x^\gamma + \left(J + \frac{K}{r} \right), \tag{92}$$

$$\frac{x}{r + \lambda - \alpha} + \nu c_1 x^\nu = \gamma k_2 x^\gamma \,, \tag{93}$$

$$\frac{y}{r + \lambda - \alpha} + c_1 y^\nu = k_2 y^\gamma + \left(\frac{K}{r} - C \right), \tag{94}$$

$$\frac{y}{r + \lambda - \alpha} + \nu c_1 y^\nu = \gamma k_2 y^\gamma \,. \tag{95}$$

These are four equations to determine four unknowns, x, y, c_1 and k_2. Suppose there exists a solution to these equations, is then the corresponding stopping rule the optimal policy?

Take the case of optimal starting of the field first. Since there is no integral term in (72), we apply theorem 1. We can pick $W = \{x \in \mathbf{R} : x > 0\}$, and the operator is clearly elliptic for $P_t > 0$. (51) is satisfied by construction, hence it only remains to prove $L_V(u(p,q) - J) \leq 0$ for $p > x(q)$. Using (78) we get

$$L_V(u - J) = rJ + L_U u + \lambda q \frac{\partial u}{\partial q}$$

$$= rJ + K - \lambda pq - \lambda \left[\frac{pq}{r + \lambda - \alpha} + \nu c_1 (pq)^\nu \right] \qquad (96)$$

$$= -\frac{r - \alpha}{r + \lambda - \alpha} pq + (rJ + K) + \lambda \nu c_1 (pq)^\nu .$$

Since $\nu < 0$ it suffices to prove:

$$\frac{r - \alpha}{r + \lambda - \alpha} pq > (rJ + K), \qquad (97)$$

and since the left hand side is increasing in p it suffices to prove this inequality for $p = x(q)$. Combining (92) and (93) we find that (97) is equivalent to

$$(r - \alpha\nu)c_1 x^\nu < (r - \alpha\gamma)k_2 x^\gamma , \qquad (98)$$

which can easily be checked for a specific solution to (92)–(95). Hence, if (98) is satisfied we can conclude that *if u is the optimal value of an open field, then v is the optimal value of a closed field.*

In the case of u, we apply theorem 2, since (71) includes an integral term. (59) is satisfied by construction, but we have to prove $L_U(v - C) \leq -f$, and to verify (60). The first inequality is treated by an argument similar to the one above. As for (60) we have that $f \geq -Ke^{-rt}$ hence by the remark following the proof of theorem 2, we conclude that (60) is satisfied.

Hence *if v is the optimal value of a closed field, then u is the optimal value of the closed field.*

To conclude, we have proved that

$$u(p,q)e^{-rt} = \sup_\tau \left\{ E^{p,q,t} \left[\int_t^\tau (\lambda Q_s P_s - K)e^{-rs} ds \right. \right.$$
$$\left. \left. + v(P_\tau, Q_\tau)e^{-r\tau} - Ce^{-r\tau} \right] \right\}, \qquad (99)$$

and

$$v(p,q)e^{-rt} = \sup_\tau \{ E^{p,q,t} [u(P_\tau, Q_\tau)e^{-r\tau} - Je^{-r\tau}] \} . \qquad (100)$$

This is similar to the optimality equation in dynamic programming. To complete the proof that u and v are optimal, an optimality equation for sequential optimal stopping is needed. This is a problem for future research.

References

Adams R.A. (1975): *Sobolev Spaces*, Academic Press.

Aronson D.G. (1967): "Bounds for the fundamental solution of a parabolic equation", *Bull AMS* 73: 890–896.

Bensoussan A., and J.L. Lions (1982): *Applications of Variational Inequalities in Stochastic Control*, North-Holland.

Bjerkholt O., and K.A. Brekke (1988): *Optimal Starting and Stopping Rules for Resource Depletion when Price is Exogenous and Stochastic*, Discussion Paper No. 40, Central Bureau of Statistics, Oslo.

Brekke, K.A. (1989): *Optimal Oil Production and Use of Oil Revenues under Uncertain Oil Prices*, thesis for the dr. polit degree, University of Oslo.

Brennan M.J., and E.S. Schwartz (1985): "Evaluating Natural Resource Investments", *Journal of Business* 58: 135–157.

Dixit A. (1989): "Entry and Exit Decisions under Uncertainty", *Journal of Political Economy* 97: 620–638.

Dynkin E.B. (1965): *Markov Processes*, Vol II, Springer-Verlag.

Friedman, A. (1976): *Stochastic Differential Equations and Applications*, Vol II, Academic Press.

Littman W., G. Stampacchia, and H.F. Weinberger (1963): "Regular points for elliptic equations with discontinuous coefficients", *Ann Scuola Norm. Sup. Pisa* 17: 43–77.

Grigolionis B.I., and A.N. Shiryayev (1966): "On Stefan's problem and optimal stopping rules for Markov processes", *Theory Prob. Applications* 11: 541–558.

Merton R.C. (1973): "Theory of rational option pricing", *Bell Journal of Economics and Management Science* 4: 141–183.

McKean H.P. (1965): "Appendix: A Free Boundary Problem for the Heat Equation Arising From a Problem of Mathematical Economics", *Industrial Management Review* 6: 32–39.

Øksendal B. (1989): *Stochastic Differential Equations*, second edition, Springer Verlag.

Øksendal B. (1990): "The High Contact Principle in Stochastic Control and Stochastic Waves", in Cinlar, Chung and Getoor (editors): *Seminar on Stochastic Processes 1989*, Birkhäuser, 177–192.

Olsen T., and G. Stensland (1988): "Optimal Shut Down Decisions in resource Extraction", *Economic Letters* 26: 215–218.

208
Kjell Arne Brekke and Bernt Øksendal

Shiryayev A.N. (1978): *Optimal Stopping Rules*, Springer-Verlag.

Samuelson P.A. (1965): "Rational Theory of Warrant Pricing", *Industrial Management Review* 6: 3–31.

Stochastic Models and Option Values
D. Lund and B. Øksendal (Editors)
© Elsevier Science Publishers B.V. (North-Holland), 1991

Invariant Controls In Stochastic Allocation Problems

Trond E. Olsen[1] and Gunnar Stensland[2]

[1] *The Norwegian Centre for*
Research in Organization and Management
N-5015 Bergen, Norway
[2] *Chr. Michelsen Institute*
N-5036 Fantoft, Norway

1. Introduction

In many resource allocation problems the rewards from different activities are correlated through a common price factor. A typical example is provided by a firm operating several mines, where the rewards from each mine are determined by the specific metal content of the ore and a common metal price. For general correlations it is normally very difficult to explicitly characterize optimal allocation policies. It is typical that the perhaps most well-known characterization result in this area, namely the Gittins index theorem for stochastic scheduling problems, relies on the assumption that rewards across activities are *not* correlated.

In this paper we identify situations in which correlation through a common price factor does *not* influence the character of optimal policies. Such circumstances are of interest because characterization results for optimal policies for independent rewards are then valid also for correlated rewards. We show that this invariance property essentially holds if the common price enters rewards multiplicatively and if the price process has a "geometric" drift component (such that the expected price is an exponential function of time).

More precisely, we consider control problems where instantaneous rewards can be written in the multiplicative form $Z_t f(X_t, u_t)$, where Z_t (the price factor) is a positive diffusion process, X_t is a vector diffusion process, possibly subject to control, and u_t is a vector of controls. We then identify two sets of conditions under which the optimal policy is "invariant" with respect to the process Z_t — in the sense that the optimal controls are independent of the realizations of Z_t: Invariance holds if

(i) Z_t has a geometric drift component and is stochastically independent of X_t, or if

(ii) both Z_t and X_t are geometric Brownian motions, possibly correlated, and X_t is not subject to control.

In both cases the optimal policy is shown to be influenced by the stochastic price factor Z_t only via simple adjustments in the parameters of the control function which is optimal for the case of a deterministic and constant price. For case (i) it is also shown that invariance in this sense will hold generally *only if* the price factor has a geometric drift component. It is worth noting that the stochastic component (the diffusion term) of the price process is of no importance in this respect, as long as the positivity requirement is satisfied.

Some interesting applications of these results are also derived. As a particular application we show how our earlier characterization (Olsen and Stensland, 1989) of an optimal policy for switching extraction between two oil fields can be generalized to the more realistic case of a stochastic oil price. In this case conditions (i) above apply. As an other application we show how an "on-off" control problem involving two geometric processes can be solved by simple parameter adjustments in the known solution (given by Dixit, 1989) for the corresponding problem involving only one process.

The latter application is an example of a problem where the reward function is linearly homogeneous in the stochastic components. The function can therefore be factorized such that one process enters rewards multiplicatively, and the invariance principle can then be applied. Special cases of this particular invariance property are known in the literature. Fischer (1978) gives a simple formula for the value of an European call option when the exercise price and the stock price follow geometric Brownian motions. Compared to Black and Scholes (1973) his formula involves the variance in the stochastic process S/X, where S is the stock price and X is the exercise price, instead of the variance in the stock price only. In addition the interest rate is adjusted for the trend in the exercise price.

McDonald and Siegel (1986) gives the value of an investment opportunity (infinitely lived American call option) where income and cost both follow geometric Brownian motions. In section 4 we show how to derive their result directly from the invariance principle.

Relative to the invariance results given (implicitly) in the literature on option pricing (Fischer 1978, McDonald-Siegel 1986), the results given here represent generalizations in the sense that they allow for more general types of stochastic processes (in particular controlled ones), and more general types of reward functions. Moreover, we elaborate on other applications.

The paper is organized as follows. The model is presented in Section 2. In Section 3 we derive the invariance results. Section 4 contains applications.

2. The model

Assume that instantaneous rewards (R) depend on a price component (z), a vector (x) describing other aspects of the allocation problem (e.g. the physical state of the system), and a vector of controls (u) in the following way: $R = zf(x, u)$. In the mining example, we could think of x as a vector describing the physical characteristics of the different mines, u as a vector describing the productive resources allocated to the various mines, and $f(x, u)$ as the resulting production rate of homogeneous metal. (We abstract from resource costs.) Assume that the state vector x can be influenced by the controls u, but that the price component Z is exogenous. Following the standard formulation of stochastic control theory, we assume that X_t and Z_t are diffusion processes. In particular, X_t is assumed to obey the (vector) stochastic differential equation

$$dX_t = b(X_t, u)dt + \sigma(X_t, u)dB_t , \qquad (2.1)$$

where B_t is a (vector) Brownian process. The price process Z_t is assumed to follow

$$dZ_t = \beta Z_t dt + \gamma(Z_t)dB'_t , \qquad (2.2)$$

where B'_t is also a Brownian process. Moreover, Z_t is assumed to be *positive* at all times. This obviously requires that the diffusion coefficient satisfies $\gamma(0) = 0$. An example of such a process is the geometric Brownian motion, where $\gamma(z) = z$.

The formulation (2.2) implies (by Dynkin's identity applied to the function $h(z) = z$) that the *expected* price grows at the constant rate β,

$$E^z Z_t = ze^{\beta t} . \qquad (2.3)$$

(Here the superscript on the expectation operator refers to the initial point $Z_0 = z$.) The property (2.3) follows from the particular form of the drift coefficient in (2.2) and will be referred to as the process Z_t having a "geometric drift component".

The objective function is assumed to be of the form

$$E^{x,z}\left[\int_o^\tau e^{-rt} Z_t f(X_t, u)\, dt + e^{-r\tau} Z_\tau g(X_\tau)\right], \qquad (2.4)$$

where τ is a stopping time w.r.t. the process X_t. More specifically, we assume that τ is either a fixed time, possibly $\tau = \infty$, or an exit time for the process X_t from some open set G. The last term in (2.4) is the terminal reward, which of course may be zero. The problem is to characterize an optimal control policy $u = U(z, x)$ such that the objective in (2.4) is maximized.

3. Results

The results below are developed under two different sets of assumptions. We first obtain an invariance result by assuming the multiplicative "price component" Z_t to be stochastically independent of the vector X_t. Then we show that similar results obtain when Z_t and X_t are correlated, provided both processes are geometric diffusions, and the processes themselves are *not* subject to controls.

A stochastically independent "price" component

Assume now that the price process Z_t is stochastically independent of the vector process X_t.

From (2.1) and (2.2) it is easily seen (e.g. from Øksendal, 1985) that the Hamilton-Jacobi-Bellman (HJB) equation for the optimal value function $H(z, x)$ takes the form

$$rH(z, x) = \sup_u \left[z f(x, u) + A^u H(z, x)\right], \quad \text{all } z, \text{ all } x \text{ in } G, \qquad (3.1)$$

where A^u, the infinitesimal operator for the process (Z_t, X_t), is a differential operator of the form

$$A^u H = \beta z \frac{\partial H}{\partial z} + \sum b_i(x, u) \frac{\partial H}{\partial x_i}$$
$$+ \sum a_{ij}(x, u) \frac{\partial^2 H}{\partial x_i \partial x_j} + \frac{1}{2}\gamma^2(z) \frac{\partial^2 H}{\partial z^2}. \qquad (3.2)$$

The supremum in (3.1) is attained for $u = U(z, x)$, where $U(\cdot)$ is the optimal control policy.

An intuitive derivation of (3.1) can be given as follows: If s is a short interval of time, the optimal value function H must obviously satisfy the dynamic programming relation

$$H(z, x) \approx \sup_u \left[z f(x, u) s + e^{-rs} E^{z,x} H(Z_s, X_s^u) \right],$$

where the superscript u refers to the fact that the process X_t is being controlled. If we subtract $e^{-rs} H(z, x)$ on both sides of this relation, then divide by s, and finally let $s \to 0$, we obtain (3.1), — by *definition* of the infinitesimal operator A^u.

Let $h(x)$ and $u(x)$ be the optimal value and control functions, respectively, for the case where the price component is constant, i.e. when $\beta = \gamma(z) = 0$. Under certain regularity conditions (see e.g. Øksendal (1985), Thm. 10.1) these functions obey the corresponding HJB equation

$$r h(x) = \sup_u \left[f(x, u) + A_0^u h(x) \right], \quad \text{all } x \text{ in } G, \tag{3.3}$$

where A_0^u is the operator obtained by setting $\beta = \gamma(z) = 0$ in (3.2), and the supremum in (3.3) is attained for $u = u(x)$. Moreover, on the boundary (∂G) of the set G we have

$$h(x) = g(x), \quad x \text{ in } \partial G. \tag{3.4}$$

In addition, we will *assume* that $h(\cdot)$ has *continuous derivatives* of first and second order on G, and that $h(\cdot)$ is continuous on the closure of G.

Now consider the problem of maximizing (2.4), given that the price process (Z_t) follows (2.2). Emphasizing the influence of the interest rate (r), let $h(x; r)$ and $u(x; r)$ denote the optimal value and control functions, respectively, corresponding to a constant price $(\beta = \gamma \equiv 0)$. It now follows directly from (3.3) that the function $H(z, x) = z h(x; r - \beta)$, satisfies the HJB equation (3.1). Moreover, from our assumptions regarding $h(\cdot)$, we see that $H(\cdot)$ is continuous on the closure of the set $G' = R^+ \cup G$, and has continuous first and second order derivatives on G'. Finally, it is clear from (3.4) that $H(\cdot)$ satisfies the boundary condition $H(z, x) = z g(x)$, all (z, x) in $\partial G'$. But then it follows from a general sufficiency theorem (Øksendal (1985), Thm. 10.2) that $H(\cdot)$ is indeed the optimal value function for problem (2.4). Moreover, since the supremum in the HJB equation (3.1) is attained for $u(x; r - \beta)$ it follows that the optimal policy is given by the function $u(x; r - \beta)$. This shows that *a price process with geometric drift exerts a very simple influence on the optimal control policy; one simply makes an adjustment of the interest rate in the policy which is optimal under a constant price*. In particular, when the (positive) price component has zero drift ($\beta = 0$) and enters rewards multiplicatively, then the optimal policy and the optimal value are both independent of the price process.

This invariance property is trivially true when the price process Z_t is deterministic and has geometric drift, i.e. when Z_t is an exponential function. The result above is thus a generalization of this property to the class of positive diffusion processes whose deterministic part (the drift component) is geometric. It shows that the exact form of the stochastic part (the diffusion component) is of no importance in this respect, as long as the positivity requirement ($Z_t > 0$) is satisfied. While easy to prove mathematically, we do not at present have a good intuitive explanation for why it is that the diffusion component has no influence on the optimal policy, once the drift component is geometric.

It is natural to ask at this point whether invariance holds for a larger class of price processes. In an appendix we show that the answer is negative, in the following sense: If the *invariance property is to hold generally* (i.e. for the class of control problems considered above) *then it is necessary that the price process has a geometric drift component*. Again this result is intuitively obvious in the deterministic case. We develop some intuition for the stochastic case in connection with the discussion of a specific application in Section 4.

Geometric processes not subject to control

Assume now that both Z_t and X_t are geometric diffusions, and that neither is subject to control. (The controls thus affect the objective function only, as in the case where an operator of a multifuel power plant must choose between fuels whose prices are out of his control). Thus, the processes Z_t, X_t have representations

$$\frac{dX_{it}}{X_{it}} = b_i dt + \sigma_{i1} dB_{1t} + \cdots + \sigma_{im} dB_{mt}, \quad i = 1, \ldots, n,$$

$$\frac{dZ_t}{Z_t} = \beta dt + \tau_1 dB_{1t} + \cdots + \tau_m dB_{mt},$$

where $(B_1 \ldots B_m)$ are independent Brownian processes. Defining the matrix $\sigma \equiv [\sigma_{ij}]$ and the vector $\tau \equiv [\tau_1 \ldots \tau_m]$, the covariance matrix is then given by

$$
\begin{bmatrix} \alpha & \phi^T \\ \phi & \gamma \end{bmatrix}, \quad \alpha \equiv \sigma\sigma^T, \quad \phi \equiv \tau\sigma^T, \quad \gamma \equiv \tau\tau^T.
$$

Here $\alpha \equiv [\alpha_{ij}]$ is the covariance matrix for the instantaneous rates of change (IRC) of the X_t-components, while $\phi \equiv [\phi_1 \ldots \phi_n]$ is the vector of covariances between dZ/Z and the IRC's of the X's.

The infinitesimal operator (3.2) now takes the form

$$
\begin{aligned}
AH = &\sum b_i x_i \frac{\partial H}{\partial x_i} + \frac{1}{2} \sum \alpha_{ij} x_i x_j \frac{\partial^2 H}{\partial x_i \partial x_j} \\
&+ \beta z \frac{\partial H}{\partial z} + \sum \phi_i x_i z \frac{\partial^2 H}{\partial x_i \partial z} + \frac{1}{2} \gamma z^2 \frac{\partial^2 H}{\partial^2 z}.
\end{aligned}
\tag{3.5}
$$

As above, consider first the case when the price component in constant, i.e. when $\beta = \gamma = \phi_i = 0$, $i = 1, \ldots, n$. Emphasizing the influence of the model parameters, let

$$
h(x; r, b, \alpha) \quad \text{and} \quad u(x; r, b, \alpha)
$$

denote the optimal value and control functions, respectively, for this case. These functions now satisfy the conditions corresponding to the HJB equation (3.3) and the boundary condition (3.4), where the operator A_0 is defined by setting $\beta = \gamma = \phi_i = 0$ in (3.5).

Then consider Z_t stochastic. First note that for any function of the form $H(z, x) = zh(x)$, the infinitesimal operator (3.5) takes the form

$$
\begin{aligned}
AH &= z \left[\sum (b_i + \phi_i) x_i \frac{\partial H}{\partial x_i} + \frac{1}{2} \sum \alpha_{ij} x_i x_j \frac{\partial^2 H}{\partial x_i \partial x_j} \right] + \beta z h \\
&\equiv z A_0' h + \beta z h,
\end{aligned}
$$

where A_0' is the infinitesimal operator corresponding to a geometric X_t-vector with drifts modified to $b_i' = b_i + \phi_i$. It then immediately follows that if we define

$$
H(z, x) = zh(x; r - \beta, b + \phi, \alpha), \tag{3.6}
$$
$$
U(z, x) = u(x; r - \beta, b + \phi, \alpha), \tag{3.7}
$$

then $H(\ )$ will satisfy the HJB equation (3.1), with the supremum attained for $u = U(z, x)$. The corresponding boundary condition will clearly also be satisfied, hence $H(\ )$ *and* $U(\ )$ *defined by (3.6–3.7) are indeed the optimal value and control functions respectively.*

In words; given the optimal value and control functions for the case when the price component Z_t is non-stochastic, the corresponding functions for the case of a stochastic price can be found by making simple adjustments in the model parameters of the former case: The interest rate is corrected for the price drift (β), while the original drift parameters (b_i) are adjusted for price correlations (ϕ_i). The optimal controls are "invariant" in the sense that they are independent of realizations of the price process Z_t; they depend only on realizations of the X_t process. In particular, if Z_t has no drift and is stochastically independent of X_t, the controls are not affected at all by the stochastic nature of the price process.

4. Applications: Stochastic scheduling problems

This section presents some implications of the results above for allocation problems of the scheduling variety, including optimal search, scheduling of R&D projects and scheduling of production activities. A famous result in this area is the Gittins index theorem, originally given in Gittins and Jones (1974) for discrete-time processes, and later extended by Karatzas (1984) to continuous-time diffusion processes. This theorem characterizes the optimal policy for a problem where a number of activities with uncertain rewards are available, only one of which can be worked upon at any moment of time. (A gambling machine with several arms fits this description, and the associated stochastic process is often referred to as the multi-armed bandit process). Given that the stochastic rewards in each activity follow independent Markov processes, the Gittins theorem states that the optimal policy is a simple reservation price (or index) policy: To each activity there can be associated an index, say $M_i(x_i)$, depending only on the parameters characterizing the reward structure on this activity (here the i'th one), and on its current state (x_i). The optimal policy is then to engage at any moment of time that activity which has the currently largest index.

An apparently critical assumption for proving optimality of the reservation price policy is that the state of every project remains *unchanged* when that project is inactive (see e.g. Whittle (1982) for the discrete-time case and Karatzas (1984) for the continuous-time case). This assumption is plainly unrealistic in many economic applications. In the mining problem for example, the unit revenues will differ across ores of different quality, but all these revenues will depend on a common product price. If this price is nonconstant over time, e.g. due to random fluctuations, then the "state" of inactive ores will *not* remain constant, and Gittins' theorem is apparently not applicable. The results above show, however, that if the "price" (a common multiplicative factor in rewards across all activities) follows a "geometric" stochastic process, then the Gittins' index policy *continues to be optimal in this situation.*

(Strictly speaking, we have shown this invariance result only for the continuous-time diffusion model, but it is straightforward to prove it directly for the discrete-time case. In the latter case, the process $Z_t, t = 0$, 1.... is said to be "geometric" if the probability distribution of Z_t, conditional on $Z_{t-1} = z$, is given by the distribution of $z \cdot \xi_t$, where the ξ_t's are i.i.d. random variables).

Optimal search

Search theory have become an important part of micro economics. The two major applications are job search and searching for the lowest price. In Stigler (1961) the problem is to find the optimal number of price observations to collect in order to minimize the sum of search costs and the price to be paid. Telser (1973) assumed that search is sequential, that is, after each observation search is continued only if the expected gain from continued search is greater than the value of stopping. Still under the assumption of sequential search Weitzman (1979) introduced different search alternatives. More recently attention has once again been focused on simultaneous search (see e.g. Gal et al. 1981).

Any *sequential* search problem can be formulated as a bandit problem, and the optimal policy is the Gittins' index policy. The Gittins' index for an alternative is similar to the reservation value in search theory. Consider a discrete-time job search problem where C_i is the search cost at alternative i and the stochastic variable X_i represents the lifetime earning in that alternative. Assuming that search is sequential, the reservation value R_i of alternative i is given by

$$R_i = -C_i + \delta E \max(X_i, R_i) \,, \qquad (4.1)$$

where δ is the discount factor (see e.g. Weitzman, 1979). The optimal policy under full recall is to examine the alternative with the greatest reservation value. Search is stopped when the maximal revealed reward is greater than the reservation value of any alternative not yet investigated. Now suppose that both search costs and lifetime wages are influenced by a common factor Z such that $C_{i,t} = Z_t \cdot C_{i,0}$ and $X_{i,t} = Z_t \cdot X'_{i,t}$. Assume Z follows a geometric process with zero trend ($E Z_t = $ constant). One could think of Z as a variable describing general fluctuations in the economy. We have shown that under such circumstances the optimal policy is independent of the realizations of Z. An important economic implication that can be drawn from this is the following: General economic growth has no impact on search behaviour if such growth is commonly believed to be just a realization of a generally fluctuating process.

This result is somewhat surprising at least for two reasons: First, we introduce a new stochastic variable which increases the variability in the wage to be accepted. Normally, increased variability increases the value of search. However, the result above shows that increased variability is not enough to influence the optimal policy as long as search costs are also influenced by the same process.

Secondly, growth increases wages which in turn increases the value of employment compared to unemployment. One might think that this would affect search behaviour. But here the value of unemployment is zero, thus growth is multiplicative on that variable too, and it follows that search behaviour is not affected. In some search models (see e.g. Jensen and Westergaard-Nielsen 1987) it is assumed that a job-seeker receives a certain unemployment insurance. Assume this insurance is influenced by economic growth in the same way as life-time wages. Thus, the term $-C_i$ in (4.1) is changed to $Z(I_i - C_i)$, where I_i is the unemployment insurance. Then increased overall wages increases the unemployment insurance, and it follows that economic growth has no impact on the optimal policy for an individual job-seeker.

Research and development (R & D)

Assume that a firm has several alternative ways to improve their present production technology (see e.g. Reinganum 1981 and Nystad and Stensland 1987). Each method needs development, but such a phase of research might be unsuccessful. If so, another method can be tried. Under the assumption that only one alternative can be investigated at a time and that the rewards between alternatives are independent, the optimal policy is to always work on the project with the greatest Gittins index. Section 3 of the present paper shows that this result can be extended to problems where all the rewards are correlated through a common stochastic multiplicative factor. This factor is best interpreted as the product price.

Optimal sequencing of production activities

Our point of departure for extending the result of Gittins and Jones is an application in optimal sequencing. Suppose we are given two petroleum reservoirs and only one platform. The production rate from each reservoir is uncertain. The problem is to allocate the production platform to the best reservoir at any time. In Olsen and Stensland (1989) we assumed a constant oil price and, using results from McDonald and Siegel (1986) and Karatzas (1984), we found a closed form expression for the Gittins index, given that the production rates follow geometric diffusions with negative trends. In this case, the index for each reservoir turned out to be of the simple form $M_i(x_i) = k_i \cdot x_i$, where x_i represents the production rate of reservoir i, and k_i is a constant determined by the parameters which characterize the stochastic production process in reservoir i. It is obviously not very realistic to assume that price is constant. The present paper shows that the Gittins index policy continues to be optimal even if it is assumed that the price follows a process with geometric drift. (However, it also shows, that this result is obviously not valid for more complex price processes.) Another restrictive assumption in this example and in the similar mining example discussed in the introduction is that there is no learning between alternatives.

The above model ignores cross effects between the two wells. In reality, such cross effects are likely to occur whenever the wells are located in the same geological structure. When producing from one well in the structure, the overall pressure will fall, and this general pressure reduction reduces production from both wells. The production effect depends on the flow characteristics of each well, which again depends on how the different parts of the reservoir are connected. A simple model which captures such cross effects is the following: Suppose the overall production potential follows a stochastic process p^1, and suppose further that the flows in the ground in the two parts of the reservoir follow processes p^2 and p^3, respectively. The production rates from wells 1 and 2 are then given by the products $(p^1 \cdot p^2)$ and $(p^1 \cdot p^3)$ respectively. Suppose further that all processes follow independent geometric diffusions, such that p_t^i is given by

$$\frac{dp_t^i}{p_t^i} = -\mu_i dt + \sigma_i dB_t^i . \qquad (4.2)$$

Here B_t^i is Brownian motion. The production rate from well $1(g_t^1 \equiv p_t^1 \cdot p_t^2)$

then follows

$$\frac{dg_t^1}{g_t^1} = -(\mu_1 + \mu_2)dt + \tau dB_t, \quad \tau = (\sigma_1^2 + \sigma_2^2)^{1/2},$$

$$B_t = \frac{(\sigma_1 B_t^1 + \sigma_2 B_t^2)}{\tau}.$$

Similarly the production rate g_t^2 from well 2 follows

$$\frac{dg_t^2}{g_t^2} = -(\mu_1 + \mu_3)dt + \tau dB_t', \quad \mu = (\sigma_1^2 + \sigma_3^2)^{1/2},$$

$$B_t' = \frac{(\sigma_1 B_t^1 + \sigma_2 B_t^3)}{\tau}.$$

It turns out that this simple construction gives a representation where the problem of which well to produce from at any stage is given by the indexes,

$$m_i = C_i \cdot \frac{p^i}{(-\mu_i + r)},$$

$$C_i = \frac{\varepsilon_i}{1 - \varepsilon_i},$$

$$\varepsilon_i = \left[\left(\frac{\mu_1}{\sigma_1^2} - \frac{1}{2} \right)^2 + \frac{2(r + \mu_i)}{\sigma_i^2} \right]^{1/2} + \frac{1}{2} - \frac{\mu_1}{\sigma_1^2}, \quad i = 2, 3.$$

This follows from the results in Olsen and Stensland (1989) where the problem is to choose the best reservoir at any time given the price is deterministic.

It follows from the analysis in Section 3 that the policy described above is optimal even if we allow for a stochastic oil price. Another obvious extention is the inclusion of proportional production cost. It is not trivial however to introduce a fixed production cost.

To illustrate the necessity of processes with geometric drift, consider the simple example above where the production equipment can be moved costlessly between two pools. Suppose that income from each pool follows a common price independent of production. If this price has a geometric drift component we show in this paper that the optimal policy is given by a simple adjustment in the interest rate of the deterministic price problem.

Now suppose this price is arithmetic with no trend. Obviously the optimal policy is changed. For negative prices, for instance, we will produce as little as possible. This underlines the importance of our positivity assumption $(Z_t > 0)$. To give some intuition for why the geometric drift component is necessary for invariance consider a sequencing problem with two pools where the product price follows

$$dZ_t = \beta dt + Z_t dB_t \, .$$

Relative to a geometric diffusion process, this process contains more information about the probability for relative price increases, in the following sense:

Given a high price p_t^H at stage t the probability for the price to rise to $p_t^H e^{r(t'-t)}$ at stage t' is small. Given a low price p_t^L at stage t the probability for the price to rise to $p_t^L e^{r(t'-t)}$ at stage t' is large. This will influence the optimal policy in the way that for low prices we will save the production from the best pool. This problem appears to have no simple solution.

To further illustrate the use of the invariance result consider the problem of chosing "on-off" control so as to maximize

$$E_0 \int_0^\infty e^{-rt} f(x_t^1, x_t^2, u) \, dt \, , \tag{4.3}$$

where $u = 1$ (on) or 0 (off) and $f(x_1, x_2, 1) = x_1 - x_2$, $f(x_1, x_2, 0) = 0$.

Here x_t^1 and x_t^2 are geometric stochastic processes, where x_t^1 and x_t^2 can be interpreted as production revenue and cost respectively, and production can be shut down and reopened without cost at will.

If we define

$$Z_t \equiv x_t^2 \, , \quad X_t \equiv \frac{x_t^1}{x_y^2} \, , \tag{4.4}$$

the objective function takes the form $z f(x, 1, u)$. Note that both Z_t and X_t are geometric processes. This problem has been solved for a non-stochastic "price" (A closed form solution for the otimal value function in this problem follows from equations 6–15 in Dixit (1989) by setting $L = k = 0$ and $P_L = P_H = w$. See also McDonald and Siegel, (1985). They use both deterministic and stochastic costs but suggest numerical integration to find the optimal value function). Let

$$h(x; r, b, \sigma^2)$$

denote the value function for this case; thus b and σ^2 denote the drift and variance, respectively, of dX/X.

The value function for the case of a stochastic "price" is then

$$H(z,x) = zh(x; r - \beta, b + \phi, \sigma^2), \quad z = x_2, \quad x = \frac{x_1}{x_2}, \quad (4.5)$$

where β and b are the drift parameters for X_2 and (X_1/X_2), respectively, ϕ is the correlation between the instantaneous rates of change (IRC) for X_2 and (X_1/X_2), and σ^2 is the variance for the IRC of (X_1/X_2). Note that $X \equiv X_1/X_2$ by Itô's lemma has the representation

$$\frac{dX}{X} = \frac{dX_1}{X_1} - \frac{dX_2}{X_2} + (\sigma_2^2 - \gamma_{12})\,dt,$$

where σ_2^2 is the variance of the IRC of X_2, and γ_{12} is the covariance of the IRC's of X_1 and X_2. In terms of the parameters of the original model (4.4), the value function (4.5) is thus given by

$$\beta = b_2,$$
$$b + \phi = b_1 - b_2,$$
$$\sigma^2 = \sigma_1^2 + \sigma_2^2 - 2\gamma_{12}. \quad (4.6)$$

Finally note that the model (4.4) can be used to solve a problem with objective function $f(x_1, x_2) = \max\{x_1, x_2\}$. This is the "maximization" variant of a multi-fuel plant operating problem.

As a final application of the "invariance principle", consider the stopping problem of McDonald and Siegel (1986)

$$\sup_\tau E_0(X_{1\tau} - X_{2\tau})e^{-r\tau}, \quad (4.7)$$

where X_1 and X_2 are correlated geometric diffusions. The solution for $X_2 = $ constant was given by Samuelson (1970). If $h(\ )$ denotes the value function for Samuelson's problem, the value function for (4.7) is then given by (4.5–4.6) above.

Appendix

It is here shown that in order for invariance to hold generally, it is necessary that the multiplicative price process has a geometric drift component. For this purpose we consider the particular class of control problems where X_t is one-dimensional and deterministic, and where $f(x, u)$ is strictly concave in u, and $b(x, u)$ is concave in u. As before, let $h(x; r)$ and $u(x; r)$ denote the optimal value and control, respectively, corresponding to $Z_t = 1$. The HJB equation is then

$$rh(x; u) = \sup_u \left[f(x, u) + b(x, u) h_x(x; r) \right],\qquad\text{(A.1)}$$

where subscripts denote derivatives. The optimal control is given by

$$f_u(x, u) + b_u(x, u) h_x(x; r) = 0.\qquad\text{(A.2)}$$

We further assume that f_u is nowhere zero in (A.2).

Next consider Z_t stochastic. Let L denote the infinitesimal operator associated with this process. The HJB equation for the value function $H(x, z)$ is now

$$rH(x, z) = \sup_u \left[zf(x, u) + b(x, u) H_x(x, z) + LH(x, z) \right].\qquad\text{(A.3)}$$

If invariance is to hold, the optimal control is of the form $u = u(x; \rho)$ for some ρ. Since the last term in (A.3) is independent of u (the Z-process is by assumption not controlled), we thus have

$$f_u(x, u) + b_u(x, u) H_x(x, z) = 0 \quad \text{for } u = u(x; \rho).\qquad\text{(A.4)}$$

From (A.2) and (A.4) it then follows that $H_x(x, z) = z h_x(x; \rho)$ which combined with the boundary condition yields $H(x, z) = z h(x; \rho)$. Now (A.1) and (A.3) yields

$$rzh(x; \rho) = 2\rho h(x; \rho) + \beta(z) h(x; \rho),$$

where $\beta(z)$ is the drift coefficient associated with Z_t. From this it follows that we must have $\beta(z) = (r - \rho)z$. **QED**

References

Dixit A. (1989): "Entry and Exit Decisions under Uncertainty", *Journal of Political Economy* 97(3): 621–638.

Black F., and M. Scholes (1973): "The pricing of options and Corporate Liabilities", *Journal of Political Economy* 81: 637–659.

Gal S., M. Landsberger, and B. Levykson (1981): "A compound strategy for search in the labor market", *International Economic Review* 22(3): 597–603.

Fischer S. (1978): "Call option pricing when the exercise price is uncertain, and the valuation of index bonds", *Journal of Finance* 33: 167–176.

Gittins J.C., and D.M. Jones (1974): "A dynamic allocation index for the sequential design of experiments", in J. Gani et al. (editors), *Progress in Statistics*, North-Holland, 241–266.

Jensen P., and W.C. Westergård-Nielsen (1987): "A search model applied to the transition from education to work", *Review of Economic Studies* 54: 461–472.

Karatzas I. (1984): "Gittins Indices in the dynamic allocation problem for diffusion processes", *Annals of Probability* 12(1): 173–192.

McDonald R., and D. Siegel (1986): "The value of waiting to invest", *Quarterly Journal of Economics* 106: 707–727.

McDonald R., and D. Siegel (1985): "Investment and the Valuation of Firms When There is an Option to Shut Down", *International Economic Review* 26: 331–349.

Nystad A., and G. Stensland (1987): "Optimal choice of R&D strategy for enhanced recovery from petroleum reservoirs", *The Energy Journal* 8(1): 125–132.

Øksendal B. (1985): *Stochastic differential equations. An introduction with applications*, Berlin: Springer Verlag.

Olsen T.E., and G. Stensland (1989): "Optimal sequencing of resource pools under uncertainty", *Journal of Environmental Economics and Management* 17: 83–99.

Reinganum J. (1981): "Dynamic games of innovation", *Journal of Economic Theory* 25: 21–41.

Samuelson P.A. (1970): "Rational Theory of Warrant Pricing", with Appendix by H.P. McKean, "Appendix: A Free Boundary Problem for the Heat Equation Arising from a Problem in Mathematical Economics", in Robert C. Merton (editor), *The collected Scientific Papers of Paul A. Samuelson*, Vol. 3, Cambridge, Mass.: MIT Press.

Stigler G.J. (1961): "The economics of information", *The Journal of Political Economy* 69: 213–225.

Telser L.G. (1973): "Searching for The Lowest Price", *American Economic Review* 63(2): 40–49.

Weitzman M.L. (1979): "Optimal search for the best alternative", *Econometrica* 47(3): 641–655.

Whittle P. (1982): *Optimization over time*, Vol. 1, New York: Whiley.

Stochastic Models and Option Values
D. Lund and B. Øksendal (Editors)
Elsevier Science Publishers B.V. (North-Holland), 1991

Shadow Prices In Stochastic Programming: Their Existence and Significance

Sjur D. Flåm

Institute of Economics
University of Bergen
N–5008 Bergen, Norway

1. Introduction

Problems of decision making under uncertainty can most often be modelled as stochastic programs of the generic form

(P) : Minimize the overall *expected* cost
$$E f(x) := E f(\omega, x_1(\omega), \ldots, x_T(\omega))$$

by making, *sequentially*, at each stage $t = 1, 2, \ldots, T$, a decision $x_t(\omega) \in \mathbf{R}^{n_t}$ under imperfect information about the exact state ω of the world. Here ω belongs to a *sample space* Ω consisting of all possible outcomes (realizations) of relevant (and exogenous) random phenomena.

   Problem (P) is constrained in *two* essential ways (i)-(ii). Most often there are

(i) **Constraints on resource endowments and technological feasibility**. They typically have the form

$$g_t(\omega, x_1(\omega), \ldots, x_t(\omega)) \leq 0 \text{ almost surely (a.s.)}$$
$$\text{for } t = 1, \ldots, T, \tag{1.1}$$

where g_t has values in $\mathbf{R}^{m_t}$, and the inequality (1.1) is understood to hold componentwise. The inequalities in (1.1) are standard in nature; they could reflect limited availability of productive resources, discrete time motion (dynamical laws), or technological possibilities. There is no need to elaborate on their precise nature here.

Written at Univ. of Bayreuth. This research has been supported in parts by Ruhrgas and NAVF.

The other, less conventional, type of constraints is related to incomplete knowledge or to uncertainty about ω. Thus there are

(ii) **Constraints on information.** These will be formalized below, (1.2). To understand them, and for the sake of intuition, it is important to keep in mind that the actual value of ω is identified only with the passage of time. Thus problem (P) incorporates a process of learning, or a stepwise production of specific knowledge. Actually, it is precisely this feature which makes (P) into a genuine sequential problem: Even when t refers to decision epochs and not directly to time, the dynamics hinge upon the stepwise resolution of uncertainty.

Apart from the value of time, one typically would like to *wait and see*, postponing, if possible, irreversible decisions (or commitments) until ω becomes better known. However, in practice there are time windows for many decisions. Some of them cannot be delayed beyond given dates, and must eventually be made *here and now*. In this manner stochastic programs bring out the distinction between *open* and *closed loop* strategies in control theory: It is unwise to commit oneself a priori to specific actions, not exploiting (or simply ignoring) the flow of information. If possible one should adapt to changes, when and if they happen, by implementing socalled feedback rules.

As mentioned, in practice, the knowledge about the specific outcome ω, in the sample space Ω, most often increases over time, or, at least, does not deteriorate. To model this nonstationary nature of uncertainty we shall naturally assume that information is generated stepwise via a *stochastic process* $\xi_1, \ldots, \xi_T$ which unfolds over time. At stage t only the values of $\xi_1, \ldots, \xi_t$ are available. Then, assuming perfect memory or recall of past values, decision x_t should depend, at most, on the actual realization of $\xi_1, \ldots, \xi_t$, and be insensitive to the (future) values of $\xi_{t+1}, \ldots, \xi_T$.

To reiterate, we insist that no decision *anticipates* (or is based upon) information which becomes available only in the future. More specifically, we could let Ω comprise all possible values of $(\xi_1, \ldots, \xi_T)$. Then formally, if $\omega = (\xi_1, \ldots, \xi_T)$ and $\omega' = (\xi_1', \ldots, \xi_T')$ satisfy $\xi_1 = \xi_1', \ldots, \xi_t = \xi_t'$, we must have $x_t(\omega) = x_t(\omega')$.

In the parlance of probability theory one says that stage t decision x_t should be Σ_t-*measurable*, where Σ_t is the sigmafield generated by $\xi_1, \ldots, \xi_t$. It is convenient to formalize this requirement as follows: At every stage $t = 1, \ldots, T$,

$$x_t = E(x_t | \xi_1, \ldots, \xi_t) \text{ a.s.}, \tag{1.2}$$

which says that each x_t should coincide with its own *conditional* expecta-tion.[1] To this we add that $g_t(\omega, \cdot)$ *should also depend only on* $\xi_1, \ldots, \xi_t$, so that the verification of (1.1) can be made at time t.

This completes the heuristic description of the multistage stochastic optimization problem. In the rest of the paper we want to discuss the two types of Lagrange multipliers (i*)-(ii*) that are associated with solutions to problem (P). They pertain to (1.1) and (1.2). To describe their nature we shall loosely say that they are related (i*) to the availability of resources (1.1) and (ii*) to information (1.2).

(i*) **Resource multipliers**

These are fairly conventional in that they "price out" the explicit (techno-logical or resource related) constraints (1.1). Indeed, they do exactly what we want and, out of old habit, expect:

- They serve to characterize optimal solutions in terms of necessary Kuhn-Tucker conditions;
- they provide keys to much of stability analysis; in particular,
- they are "shadow prices" that help to evaluate marginal relaxations of constraints; moreover,
- they play major roles in methods involving duality or decomposition; and, also important,
- they are central to the design of exact penalty functions.

These benefits of standard multipliers related to (1.1) are fairly well known. It is less known that *with stochasticity present, these (resource) multipliers take on new features; they become* either *random vectors*, or *they are measures*. This has all been clearly brought to the fore in a series of papers by Rockafellar and Wets (1975), (1976), (1977), (1978), (1983).

What confronts us here is parts of functional analysis.[2] Not surprisingly, since problem (P) admits strategies $x_t(\omega)$ depending on random data ω, we need to deal with various spaces of measurable functions. In this regard we feel that parts of stochastic programming have sometimes embarked on routes with difficult, exclusive or narrow access.[3] Quite naturally there has been a preference for employing functions spaces on which it is fairly

---

[1]   Note that (1.2) is a basically *linear*, and thus, in principle, tractable constraint whereas (1.1) may generally be highly nonlinear.

[2]   For good references see Holmes (1975), or Kantorovich and Akilov (1982).

[3]   The same objection applies to segments of mathematical economics dealing with problems involving infinite horizon or inifinitely many commodities, see e.g. Prescott and Lucas (1972).

easy to check regularity assumptions (continuity, differentiability etc.), and straightforward to verify constraint qualifications (say e.g. strict feasibility). However, as a result, one has been obliged to accept multipliers from nontractable (*dual*) spaces which do not accord with economic intuition.

The standard example is $L^\infty(\Omega)$, the space of essentially bounded measurable function on Ω, whose dual comprises socalled singular elements. Broadly speaking, the problem here is that if the original space is "too small", as e.g. $L^\infty(\Omega)$, then its dual space is "too large". To avoid intricate problems of this type, and to follow Aristotle in his quest for a middle avenue, we shall strike a balance, admitting rather large, yet natural and pleasant spaces whose duals are equally tractable. In that endeavour we shall complement the studies of Rockafellar and Wets by focusing on "nicer" spaces than $L^\infty(\Omega)$ and $L^1(\Omega)$. Thus the works of Eisner and Olsen (1974), and Wets (1970) may be regarded as predecessors of this paper.

Also, we shall point out, at the appropriate junction, that the choice of function space is a *modelling* issue which should not be completely relegated as a mathematical technicality.

(ii*) **Multipliers on information**

Guided by experience in deterministic programming one would naturally strive for decoupling not only (1.1) but also the informational constraints (1.2) from problem (P) by means of Lagrange multipliers. In fact, this project is often feasible as made clear in many studies, see Rockafellar and Wets (1976), Dempster (1981), Flåm (1985). Moreover, as explained by Evstigneev (1976), (1985), the relevant multipliers yield a rich interpretation in terms of insurance.

A major objective of this paper is to convey that in stochastic programming one can use the techniques of Lagrangian duality and relaxation pretty much along familiar lines. The important thing to note is that multipliers then become random variables.

This may all leave the *false* impression that there is nothing essentially new or peculiar to this branch of optimization. Quite to the contrary, stochastic programming is very challenging in all directions: modelling, analysis and computation, see Varayia and Wets (1988). Our excuses for not touching upon these issues are threefold. First, modelling issues are better left to specific problems. Second, this paper is *not* aimed at a mathematical analysis of problem (P). That analysis constitutes a separate paper, Flåm (forthcoming). Third, typical instances of problem (P) are computationally very hard, requiring special techniques of decomposition.

Even linear, two-stage versions can often be solved only approximately, see Birge and Wallace (1988) and Kall (1988). Computational concerns are however, beyond the scope of this paper, which is organized as follows.

Section 2 provides preliminaries for the subsequent discussion.

Section 3 and 4 deal with multipliers associated to constraints on resources (1.1) and information (1.2), respectively.

2. Assumptions

This section lists the technical hypotheses that we shall impose on (P). The presentation is designed to avoid mathematical details, but can be skipped on first reading or by less technically oriented readers.

It is appropriate to recall, first of all, that optimization problems typically involve *three* spaces: One for decisions, one for perturbations of the given data, and one for multipliers. We shall discuss all these in that order.

On strategy spaces

Recall that the decision $x_t(\xi_1, \ldots, \xi_t)$ at stage t should be a (Borel measurable) function of all information $\xi_1, \ldots, \xi_t$ available to date. More generally, we can require that x_t belongs to some *subspace* X_t of all Σ_t-*measurable* random vectors in $\mathbf{R}^{n_t}$, where Σ_t is the sigma-field generated by $\xi_1, \ldots, \xi_t$.[4] The precise, and problem specific, nature of X_t will not be our immediate concern. However, when in Section 4 we come to prices on information, we shall need that X_t admits a dual space P_t of multipliers p_t which can all be represented in terms of expectation (i.e. as integrals)

$$E(p_t \cdot x_t | \xi_1, \ldots, \xi_t) \,.$$

On spaces of perturbations

The standard (and canonical) way to perturb (P) is to replace (1.1) by

$$g_t(\omega, x_1(\omega), \ldots, x_t(\omega)) \leq u_t(\omega) \text{ a.s. for } t = 1, \ldots, T \,, \tag{2.1}$$

---

[4] We shall here leave x_t at this, but often it is natural to go somewhat further by requiring that X_t be a *Banach subspace* of all $\mathbf{R}^{n_t}$-valued, Σ_t-measurable random vectors. Then, we could adduce to problem (P) a constraint $x \in C$, where $C \subset X_1 \times \ldots \times X_T$ is closed, nonempty, "simple" set, intentionally kept outside the process of "Lagrange multiplication". But we shall avoid to burden the arguments with these things which are not essential. For details see Flåm (forthcoming).

where the *perturbation* $u_t(\omega) \in \mathbf{R}^{m_t}$ on the right hand side of (2.1) belongs to some subspace U_t of measurable random vectors. Here, if $u_t(\omega) \geq 0$ (in all coordinates), it can be interpreted as additional, random resources yielding a relaxation of problem (P).

A crucial question is: what type of perturbations u_t should be considered? In this respect we shall be guided by one desire; namely: *For duality we need to evaluate perturbations by means of continuous linear price regimes.* Then such regimes should come in a handy and intuitively understandable form. Specifically, *we always want price operators to be representable* as *integral operators*, i.e., given prices y_t, the perturbation u_t should cost all together

$$y_t(u_t) := <y_t, u_t> := E(y_t(\omega) \cdot u_t(\omega)).$$

In terms of functional analysis this means that the *dual space* Y_t, consisting of all continuous linear functionals on U_t, *should admit an integral representation for each of its elements.* To increase chances for having unique prices, we shall also require that *the unit ball of the price space be strictly convex.*[5]

There is, of course, an intimate relation between decision spaces and acceptable perturbations. Indeed, what we need is that

$$x_1 \epsilon X_1, \ldots, x_t \epsilon X_t \to g_t(\cdot, x_1(\cdot), \ldots, x_t(\cdot)) \in U_t \quad \text{all } t. \qquad (2.2)$$

Having tailored the spaces $X_t, U_t, t = 1, \ldots, T$, to each other so that (2.2) holds, we are finally left with no freedom concerning

The space of resource multipliers

These are now bound to coincide with the dual spaces $Y_1, \ldots, Y_T$ of $U_1, \ldots, U_T$, respectively.

Clearly, we cannot hope to get away without imposing a few

---

[5] Examples include the $L^p(\Omega)$ spaces, $1 < p < +\infty$, and their generalization to Orliz spaces where the $\triangle_2$-condition holds. Also, when Ω is countable, the spaces $L^p(\Omega)$, $1 < p < +\infty$, would be acceptable.

   Important properties of these spaces are first, that *convergence in norm implies convergence in probability;* second, that *the dual space* Y_t *of* U_t *has strictly convex unit ball,* and third, that *its elements* $y_t \in Y_t$, *which by definition are continuous linear functionals on* U_t, *are representable as integrands*, i.e., $y_t(u_t) = E[y_t(\omega) \cdot u_t(\omega)]$ for some random vector $y_t \in \mathbf{R}^{m_t}$. For details see Kantorovich and Akilov (1982).

Regularity assumptions on the problem data.

In this regard we shall assume that the criterion $Ef(\omega, x(\omega))$ *is locally Lipschitz.* For example, if $X := X_1 \times \ldots \times X_T$ is a closed subspace of L^∞ and

$$|f(\omega, x) - f(\omega, x')| \leq k(\omega)|x - x'|$$

for every pair of *constant* vectors $x, x' \in X$, with $k \in L^1$, then the above assumption holds with Lipschitz constant $E(k)$, see Hiriart-Urruty (1978, Lemma 3). We also assume *that each*

$$g_t : X_1 \times \ldots \times X_t \to U_t, \quad t = 1, \ldots, T,$$

is locally Lipschitz continuous.

To all of this we add that a few innocous assumptions on the probability space Ω.[6]

3. The multiplier rule for ordinary constraints

This section focuses exclusively on multipliers associated with the explicit constraints (1.1). Only here do we present some novelties.

It is efficient to organize the presentation around the duality scheme of Rockafellar (1974). For the sake of the argument, suppose the perturbation $u_t \in U_t$ in (2.1) can be bought only at an extra expense

$$y_t(u_t) := <y_t, u_t> := E(y_t \cdot u_t).$$

Here y_t belongs to the dual space Y_t of U_t, and should be interpreted as a random, exogenous price regime revealed only at stage t. In broad terms, duality is concerned with stipulating these prices. Before adressing that issue it is best to emphasize that y_t is a random vector, and to explain in what

---

[6] We shall require that all σ-algebras $\Sigma_1, \ldots, \Sigma_T$ be *complete*. If some Σ_t is not, it should be completed in the standard way. This being done, we identify measurable functions (or events) that coincide a.s.

We also demand that $\Sigma_1, \ldots, \Sigma_T$ be *separable* in the sense that each Σ_t becomes a separable metric space when endowed with the distance function

$$d(A, B) := \Pr(A \backslash B) + \Pr(B \backslash A).$$

sense. For this purpose recall that at stage t, only the values of $\xi_1, \ldots, \xi_t$, are known. That is to say, to decide whether (1.1) holds or not, we need (only) be informed about the realizations of $\xi_1, \ldots, \xi_t$. Thus when evaluating desirable perturbations u_t, it is impossible to discriminate between different candidates with finer precision than allowed for by $\xi_1, \ldots, \xi_t$.

This amounts to have the perturbation u_t itself be a (Borel measurable) function of $\xi_1, \ldots, \xi_t$.[7] The price y_t, unveiled in period t and prior to the choice of x_t, should, for the same reason, also be a (Borel measurable) function of the currently available information $\xi_1, \ldots, \xi_t$.

In summary, the opportunity to procure ourselves with perturbation profiles

$$u := (u_1, \ldots, u_T) \in U_1 \times \ldots \times U_T$$

at the additional cost

$$<y, u> := <y_1, u_1> + \cdots + <y_T, u_T>,$$

leads naturally to the *Lagrangian*

$$L(x, y) := \inf_u \{ Ef(x) + <y, u> \,|\, g_t(x_1, \ldots, x_t) \leq u_t \text{ a.s.}, \quad t = 1, \ldots, T \}.$$

Trivially, $L(x, y) = -\infty$ if for some t, and $u_t \geq 0$ a.s., we have $<y_t, u_t> < 0$. Therefore, only non-negative prices $y_t \geq 0$ a.s. are worthy of further consideration, and then the Lagrangian takes on the familiar form

$$L(x, y) = Ef(x) + \sum_{t=1}^{T} <y_t, g_t(x_1, \ldots, x_t)> .$$

The original (*primal*) problem (P) can be compactly restated as

$$(P): \inf_{x \in C} \sup_{y \in Y_+} L(x, y),$$

where Y_+ denotes the non-negative cone of $Y = Y_1 \times \ldots \times Y_T$. Its value $\inf(P)$ majorizes that of the associate *dual problem*

$$(D): \sup_{y \in Y_+} \inf_{x \in C} L(x, y).$$

---

[7] Which equivalently means Σ_t-measurable.

A saddle point (x, y) of L would solve (P) and (D) optimally with equal values, and must necessarily satisfy the *Kuhn-Tucker conditions*

$$0 = <y, g(x)> \qquad (3.1)$$

$$0 \in \partial_x L(x, y), \qquad (3.2)$$

where ∂_x denotes the partial subdifferential of convex analysis. y is then called a *Lagrange multiplier* at x.

The analysis of Rockafellar and Wets is confined to the case when $f(\omega, \cdot)$ and each component of $g_t(\omega, \cdot)$ is convex, with special attention on conditions which ensure existence of primal-dual optimal pairs of solutions. Their approach is the min-max theory of the Lagrangian saddlefunction introduced here above.

By contrast, we shall dispense with convexity assumptions, and take existence of a locally optimal solution to (P) for granted. Our main concern is to identify the precise nature of multipliers, to guarantee their existence, and finally, to provide constraint qualifications which imply that the multiplier rule has the desired normal form.[8] If problem (P) is non-convex, it is often unrealistic to search for global saddle points. Rather we may have to contend, at best, with local versions of such points. Yet, it is not wishful thinking to maintain (3.1–2). In effect, relaxing on convexity, we can still accomodate for lack of smoothness with ∂_x signifying then the Clarke (1983) subdifferential. Readers unfamiliar with this generalized calculus should simply interprete ∂_x as the usual operation of taking the partial derivative.

The following result attests that *local* solutions to (P) are supported by Lagrange multipliers.

Theorem 3.1. (Fritz John multiplier rule)
Under the hypothesis of Section 2 suppose each space U_t of perturbations is separable, and let x be a locally optimal solution to problem (P). Then there exists a non-zero, non-negative multiplier

$$y = (y_0, y_1, \ldots, y_T) \in R \times Y_1 \times \ldots \times Y_T$$

---

[8] The vehichle used in Flåm (forthcoming) is the non-smooth calculus of Clarke (1983). Hiriart-Urruty (1978) has already applied non-smooth calculus to 2–stage versions of our problem. He prefered to treat (P) in its dynamic form, and did not explore to what function space the second stage multipliers belong. By contrast, we avoid dynamic programming and rather analyze (P) in its extended "static" form. The merits of this approach is that a multiplier rule emerges that completely parallels familiar results for deterministic programs.

such that

$$y_t(\omega) \cdot g_t(\omega, x_1(\omega), \ldots, x_t(\omega)) = 0 \quad \text{a.s. for } t = 1, \ldots, T, \quad (3.3)$$

and

$$0 \in \partial_x \left[y_0 E f(x) + \sum_{t=1}^{T} < y_t, g_t(x_1, \ldots, x_t) > \right]. \quad (3.4)$$

For a proof see Flåm (forthcoming). As in Clarke (1983, Thm. 2.7.5) conditions can be given which ensure that (3.4) holds a.s., that is to say pointwise:

$$0 \in \partial_x \left[y_0 E f(\omega, x(\omega)) + \sum_{t=1}^{T} y_t(\omega) g_t(\omega, x_1(\omega), \ldots, x_t(\omega)) \right] \text{ a.s.}$$

For numerous purposes it is convenient that the only multiplier $y = (y_0, y_1, \ldots, y_T)$ satisfying (3.3–4) with $y_0 = 0$ is the origin 0. We shall then say that the *multiplier rule holds in normal Lagrange form*. In smooth, deterministic programming a minimal hypothesis toward this is (equivalent to) the well known Mangasarian-Fromowitz (M-F) constraint qualification. If, moreover, data are convex with equality constraints being absent, then the M-F-qualification reduces to the Slater condition. To conclude this section we next mention that this all carries over to the stochastic program (P).

For a precise statement we begin with the most general *M-F-constraint qualification*. According to this there should exist a random direction $d = (d_1, \ldots, d_T) \in X$ such that for $t = 1, \ldots, T$, we have

$$g_t(\omega, x_1(\omega), \ldots, x_t(\omega)) = 0 \rightarrow g_t^0(\omega, x_1(\omega), \ldots, x_t(\omega);$$
$$d_1(\omega), \ldots, d_t(\omega)) < 0 \text{ a.s.}$$

Here g_t^0 denotes, for each given ω, the *Clarke directional derivative* of $g_t(\omega, x_1(\omega), \ldots, x_t(\omega))$ at the point $(x_1(\omega), \ldots, x_t(\omega))$ in the direction $(d_1(\omega), \ldots, d_t(\omega))$.[9]

Proposition. Under the M-F constraint qualification all multipliers at the local solution x to P have normal form. This qualification holds if there exist some $\hat{x}$ such that

$$g_t(\omega, \hat{x}_1(\omega), \ldots, \hat{x}_t(\omega)) < 0 \quad \text{a.s. for } t = 1, \ldots, T,$$

with $g_1, \ldots, g_T$ being starshaped with respect to $\hat{x}$.

---

[9] Again readers unfamiliar with this notion should simply interprete g_t^0 as a usual directional derivative.

4. Multipliers on information

So far we have treated (1.2) as an *implicit* constraint in problem (P), meaning that the (essential) objective of (P) equals $+\infty$ if (1.2) is violated at some stage.

Now it is time to treat (1.2) *explicitly*, and to relax (P) by means of *price regimes* p_t, $t = 1, \ldots, T$, *on information*. To see precisely how, we rely once again on Rockafellar's scheme of conjugate duality (1974), perturbing now (1.2) to

$$x_t = E(x_t | \xi_1, \ldots, \xi_t) + v_t, \tag{1.2'}$$

where the (random) perturbation v_t (and thus x_t) is allowed to depend on *future* information, i.e., v_t can be a function of $\xi_1, \ldots, \xi_t$ *and* some, possibly all, of the variates $\xi_{1+t}, \ldots, \xi_T$ as well. Formally, we have relaxed the constraint $x_t \in X_t$ to $x_t \in \chi_t$, where χ_t is a larger space than (contains) X_t. Suppose that such perturbations v_t are available at fixed (unit) prices p_t net of the expense $E(p_t \cdot E(x_t | \xi_1, \ldots, \xi_t))$. Here the multiplier p_t is a continuous linear functional on χ_t. Following the approch here above let us assume that any such p_t can be represented as an expectation, i.e.,

$$p_t(x_t) := E(p_t \cdot x_t).$$

The Lagrangian, which now takes on the form

$$L = E f(x) + \Sigma_t E(y_t \cdot g_t + p_t \cdot x_t),$$

should be maximal in p_t at every local solution x to (P). Using

$$E(\cdot) = E[E(\cdot | \xi_1, \ldots, \xi_t)],$$

and

$$E(p_t \cdot E(x_t | \xi_1, \ldots, \xi_t)) = E(p_t | \xi_1, \ldots, \xi_t) \cdot E(x_t | \xi_1, \ldots, \xi_t),$$

we see that the supremum of L with respect to p_t would equal $+\infty$, unless the following *complementarity condition* holds:

$$E(p_t | \xi_1, \ldots, \xi_t) = 0 \text{ a.s.} \tag{4.1}$$

Thus the *conditional mean value of information is nil* (subject to decisions being nonanticipative). We stress that p_t is a random vector of the same dimension as x_t. However, and this is important, p_t depends on *more information* than just $\xi_1, \ldots, \xi_t$. This distinguishes p_t from the (resource) multiplier y_t associated with (1.1). The probabilistic condition (4.1) can be cast in the equivalent form

$$E(p_t \cdot x_t) = 0 \quad \text{for all decisions } x_t \in X_t.$$

Geometrically the latter form means that p_t must be orthogonal to the linear subspace of functions satisfying (1.2). It turns out that p_t can be interpreted as a *net insurance compensation*, that is to say, paying an *insurance premium* $\pi_t \cdot x_t$ at stage t and receiving $(p_t + \pi_t) \cdot x_t$ as compensation later on, relieves us (in expectation and at stage t) of the burden of uncertainty. For more about this see Back and Pliska (1986), Dempster (1981), Evstigneev (1976), (1985), Flåm (1985), Rockafellar and Wets (1976).

Computationally, and conceptually, the knowledge of $p_1, \ldots, p_T$ would be most beneficial: Given these functions we could solve problem (P) "out-comewise", i.e., for *almost all realizations of* $\xi_1, \ldots, \xi_T$ *solve the* corresponding *deterministic* problem (detP): minimize

$$f(\xi_1, \ldots, \xi_T, x_1, \ldots, x_T) + \Sigma_t p_t(\xi_1, \ldots, \xi_t) \cdot x_t$$
$$\text{subject to } g_t(\xi_1, \ldots, \xi_t, x_1, \ldots x_t) \leq 0, t = 1, \ldots, T.$$

The resulting solution $x = (x_1, \ldots, x_T)$ depends parametrically on $\xi_1, \ldots, \xi_T$, and can, if all functions $f, g_1, \ldots, g_T$ are convex in x, be chosen as optimal *strategies* for the original problem (P).

5. Concluding remarks

A tenable, but limited view on stochastic programming goes as follows: In the final round, when we actually come to real computations, any probability space will be represented as a *finite* collection of possible outcomes. Then, for practical purposes, all function spaces of (measurable) strategies are finitedimensional, and functional analysis can be totally ignored. Thus, it seems that essential difficulties only originate with large scales, and with the curse of dimensionality. I find this perspective too narrow in several ways:

- It keeps us away from identifying appropriate spaces of decisions, perturbations and multipliers;
- It tends to forget issues concerning the quality of approximations involving discrete probability measures;

- It overlooks the theoretical efficiency of calculus;
- It does not acknowledge that to "compute" expectations is *conceptually a very simple operation*;
- It does not satisfy a real need for a (mathematical) analysis of stochatic programs.

It has been an aim of this paper to show that very often this analysis can be done along lines already well known within determinstic programming.

References

Back K., and S.R. Pliska (1987): "The shadow price of information in continuous time decision models", *Stochastics* 22: 151–186.

Birge J.R., and S.W. Wallace (1988): "A separable piecewise linear upper bound for stochastic linear programs", *SIAM J. Control and Opimization* 26(3): 725–739.

Clarke F.H. (1983): *Optimization and Nonsmooth Analysis*, New York: J. Wiley & Sons.

Dempster M.A.H. (1981): "The expected value of perfect information in the optimal evolution of stochastic systems", Working Paper, IIASA 55, appeared in M. Arato and A.V. Balakrishnan (editors), *Stochastic Differential Systems; Proceedings of the IFIP Working Conference, Visigrad, Hungary 1980*, Springer Verlag Lecture Notes in Control and Information Sciences.

Eisner M., and P. Olsen (1974): "Duality for stochastic programming, interpreted as L. P. in L_p-space", *SIAM J. Appl. Math.* 28: 779–792.

Evstigneev I.V. (1985): "Prices on information and stochastic insurance models", Working paper CP 23, IIASA.

Evstigneev I.V. (1976): "Lagrange multipliers for problems of stochastic programming", *Lecture Notes Econ. Math. Systems* 133.

Flåm S.D. (1985): "Nonanticipativity in stochastic programming", *J. of Optimization Theory and Applications* 46(1): 23–30.

Flåm S.D. (forthcoming): "Lagrange multipliers in stochastic programming", to appear in *SIAM J. of Control and Optimization*.

Hiriart-Urruty J.-B. (1978): "Conditions necessaires d'optimalite pour un programme stochastique avec recours", *SIAM J. Control and Optimization* 16: 317–329.

Holmes R.B. (1975): *Geometric Functional Analyses and its Applications*, Berlin: Springer Verlag.

Kall P. (1988): "Stochastic programming with recourse. Upper bounds and moment problems", *Mathematical Research*, Berlin: Acad. Verlag, 45: 86–103.

Kantorovich L.V., and G.P. Akilov (1982): *Functional Analysis*, Oxford: Pergamon Press.

Prescott E.C., and R.E. Lucas, jr. (1972): "A note on price systems in infinite dimensional space", *Int. Econ. Review* 13: 416–422.

Rockafellar R.T. (1974): *Conjugate Duality and Optimization*, Conference Board of Math. Sciences Ser. No. 16, SIAM, Philadephia.

Rockafellar R.T., and R.J.-B. Wets (1975): "Stochastic convex programming: KuhnTucker conditions", *J. Mathem. Economics* 2: 349–370.

Rockafellar R.T., and R.J.-B. Wets (1976): "Nonanticipativity and L^1-martingales in stochastic optimization problems", *Math. Programming Studies* 6: 170–187.

Rockafellar R.T., and R.J.-B. Wets (1977): "Measures as Lagrange multipliers in multistage stochastic programming", *J. of Math. Anal. and Applic.* 60: 301–313.

Rockafellar R.T., and R.J.-B. Wets (1978): "The optimal recourse problem in discrete time", *SIAM J. Control and Optimization* 16: 16–36.

Rockafellar R.T., and R.J.-B. Wets (1983): "Deterministic and stochastic optimization problems of Bolza type in discrete time", *Stochastics* 10: 273–312.

Varaiya P., and R.J.-B Wets (1988): "Stochastic dynamic optimization; Approaches and computation", Working paper, 88–87 IIASA.

Wets R.J.-B. (1970): "Problèmes duaux en programmation stochastique", *C.R. Acad. Sci. Paris*, Ser. AB 270: 47–50.

IV. Statistical Models of
Natural Resource Exploitation

Stochastic Models and Option Values
D. Lund and B. Øksendal (Editors)
© Elsevier Science Publishers B.V. (North-Holland), 1991

Estimating Structural Resource Models When Stock Is Uncertain: Theory and Its Application to Pacific Halibut

Peter Berck[1] and Grace Johns[2]

[1]*University of California at Berkeley, CA 94720*
[2]*Spectrum Economics, Inc., San Francisco, California*

Economists and policymakers frequently face the problem of making decisions about stochastic systems but oftentimes do not directly observe the most important elements. Common examples include managing fisheries when one does not directly observe the stock of fish, making energy policy when unproved reserves are not known, restricting immigration without a clear knowledge of the number of undocumented aliens, and designing a policy to fight drug addiction when the supply of drugs is not observable. All of these policy problems have an unobservable component which is critical to understanding the behavior of the system being studied. A combination of the methods of maximum likelihood and the Kalman filter provides a way to estimate the parameters of the stochastic difference equations that govern the evolution of resource stocks. Much of the problem of regulating fisheries stems from the great variance in the fish stock from season to season. Environmental factors (such as water temperature and fishing) explain some of the apparent changes in year-to-year fish stocks, but a good deal of the variance cannot be explained by deterministic means and is, therefore, taken as stochastic.

In the next section of this paper, we set out the theory of the Kalman filter/maximum likelihood method of estimation in a simple fishery setting. We contrast our view with the more common estimation practice. Section 3 of the paper applies these ideas to a real fishery problem, the Pacific Halibut fishery. The model in section 3 is a good deal more complex than that in the second section. The final section includes some conclusions about the

The authors would like to thank R.T. Carson, D. Brillinger, S. Flåm and L. Breiman for their help and suggestions during the model development and the general implementation of some of the aspects of this paper. The authors accept responsibility for any errors.

halibut example and some suggestions for other areas in agricultural and resource economics where application of the filter idea can enhance stock estimates and thus policy decisions.

The maximum likelihood Kalman filter

Since one cannot observe the actual stock of fish in the sea, it is natural to use a statistic one can observe as a proxy. The proxy that should be chosen depends upon the regulation of the fishery. We will begin this section with a presentation of a simple, dynamic fishery model. The production function in the model is chosen so that yield-per-unit effort is an indicator of the fish stock. That model is then expanded to include a quota on total fishery harvest. When the quota is harvested, the season is over and the season length is measured in days. In this regulated model, it is catch-per-day-per-unit effort that proxies the unobserved stock. Since most fisheries are regulated, these changes are not trivial.

The simplest model of an unregulated, open access fishery determines the (stochastic) time path of fishing effort, E; catch, h; and fish, x, as functions of (exogenously) given prices for fish, p, and costs of maintaining one unit of effort, c.[1] The simplest model, which is to say a too simple model, begins with a law, $F()$, governing the growth of the fish stock. The model ignores age classes, predators, etc., to achieve this simplicity.

$$x_{t+1} = F(x_t - h_t, \beta) + w_t, \tag{1}$$

where w is a normally distributed mean zero variate with variance W, β is a parameter vector, and t is time in years. All the additive error terms in this paper are assumed uncorrelated with each other and are assumed to be serially uncorrelated.

The second equation of this model is a standard equation of yield-effort models,

$$h_t = kE_t x_t + E_t v_{yt}, \quad \text{or}$$
$$y_t = \frac{h_t}{E_t} = kx_t + v_{yt}. \tag{2}$$

Equation (2) states that yield-per-unit effort, y, is proportional to stock with constant of proportionality, k. The random error in the first form

---

[1] The model generalizes to n-dimensions. In all that follows, one could interpret x as a vector of state variables and y as a vector of measurement variables. All the arguments remain the same.

of equation (2) has variance proportional to effort. In the second form of the equation, the random error, v_{yt}, is distributed normal with mean zero and variance, V_y. Equation (2) is called a measurement equation because the dependent variables are observable and depend on the unobservable stock, x. Equation (2) is just a production function giving output, h, as a function of a single input, E, and an uncontrollable variable, x, that plays a role formally equivalent to technical progress. Again, an additive normal error is assumed. Any other production function that maintains the role of x would also work, assuming one knew the coefficients. Finally, annual effort is modeled as

$$E_{t+1} = E_t + \delta(py_t - c)E_t + v_{et}, \tag{3}$$

where v_{et} is a random error term. This last equation embodies the notion that positive profits per unit effort (the term inside parentheses) lead to entry of effort at rate δ. Negative profits lead to a decrease in effort. It is the naive long-run model of the firm: Positive profits mean (slow) entry of firms and effort. A more elaborate version of the simple model would come from modeling the components of effort: boats and effort per boat and allowing the components to vary. One would then amend equation(2) so that price equals marginal cost. As it stands, an interpretation is that each firm finds it profitable to fish exactly 1 unit of effort which is presumably equivalent to the maximum amount they can fish. Smith (1968) is usually credited with the three-equation dynamic version of this model. His model is more elaborate and allows for crowding and for differing production functions. Obviously, it does not matter whether one uses the primal $h(E)$ point of view or the dual $c(h)$ formalism, though there is certainly fishery literature that argues this choice is meaningful (Fullenbaum, Carlson, and Bell, 1971). In short, this is the simplest dynamic model that could be called a "fishery".

When a fishery is regulated, a slightly different model is appropriate. Abstracting quite a bit, Pacific Halibut are regulated by means of a quota on total catch of the fishery, q. The quota is enforced by closing the fishery when the quota is met. The fraction of the potential fishing season during which harvesting is allowed is called σ. We shall make the extreme assumption that, as soon as the fishery is closed, the boats just sit in port earning nothing. Equation (1) is unchanged by this regulation since it is merely a statement of the biology. Equation (2) needs some work:

$$q = h = kEx\sigma + v_y E\sigma, \quad \text{or}$$
$$y = \frac{h}{E\sigma} = kx + v_y, \tag{2r}$$

modifies equation (2) by taking the effective effort as $E\sigma$ or effort times the percent of the season it is used. The adjustment equation for effort depends upon per boat profits. Regulation leaves costs unaffected, but per boat revenues are now (σpy), so

$$E_{t+1} = E_t \delta(\sigma_t py_t - c)E_t + v_{et}. \tag{3r}$$

Equations (1), (2r), and (3r) are the simple regulated fishery. The entry equation (3r) is particularly naive. It assumes that firms act in accord with instantaneous profits and not rational expectations about their present value of profits (Berck and Perloff, 1984).

Typically, one observes all the variables except stock, x, and wishes to estimate the parameters, δ and k, and whatever parameters are in the biomass size function, F. Also, one should be estimating a regulated model, rather than the more popular unregulated one, because fisheries are regulated.

Estimation with unknown stock

The parameters of the equations representing the simple fishery and the fish stock can be estimated by a combination of the Kalman filter and maximum likelihood. In this section, we will explain how this procedure is done in simplified terms and for the simplified model. We follow the presentation in Meinhold and Singpurwalla (1983). More standard and more detailed explications can be found in Harvey (1981) or Gelb (1974). The models presented in this section differ from the estimated models primarily in the number of measurement equations, the choice of additive rather than multiplicative errors, and the need for an extended filter in the empirical work.

Equation (1) is called a state equation because it describes the evolution of the unobservable state variable, fish stock. Its value is never known. The best estimate of stock at time t, given all observations (on y and h) up to and including $t - 1$, is denoted $x_{t|t-1}$. This is shorthand for a normal random variate with mean $\hat{x}_{t|t-1}$ and variance $P_{t|t-1}$. One then observes y_t. This new information leads to a revised estimate of the fish stock. This new best estimate, given information through time t, is denoted $x_{t|t}$; and it is again a normal random variate with mean $\hat{x}_{t|t}$ and variance $P_{t|t}$. Given β, W, and V_y, or estimates of them, the Kalman filter is an algorithm to determine $x_{t|t}$ when one knows $x_{t-1|t-1}$ and realizations of the variables, y_t, which depend on x_t. The filter also gives the variance of $x_{t|t}$, denoted $P_{t|t}$, which is determined from $P_{t-1|t-1}$ and y_t.

For the simplest case, which is all we shall describe here, assume that $\hat{x}_0$ and $P_{0|0}$ are the mean and variance of x at time zero.[2] To keep the example simple, also assume that the stock function, F, in equation (1) is linear. The filter will give an estimate $x_{1|1}$ in terms of $x_{0|0}$ and y_1. Similarly, one gets estimates for time 2 from those of time 1 and so on. Therefore, it suffices to consider the general case of getting $x_{t|t}$ from $x_{t-1|t-1}$.

At time $t - 1$, which is to say before y_t is observed, the estimate of x_t, called $x_{t|t-1}$, is $F(x_{t-1|t-1} - h_{t-1}, \beta)$. This estimate is a normal random variable because it is a linear function of a normal random variable and some constants. To be explicit, let $x_{t|t-1} = F + w = \beta_0 + \beta_1(x_{t-1|t-1} - h_{t-1}) + w$. Its mean is $\beta_0 + \beta_1(\hat{x}_{t-1|t-1} - h_{t-1})$. Its variance is the variance of w, called W, plus $\beta_1 P_{t-1|t-1}\beta_1'$. The latter term is the contribution of the randomness of $x_{t-1|t-1}$ to the randomness of $x_{t|t-1}$. For later use, let this variance be called

$$R_t = P_{t|t-1} = \beta_1 P_{t-1|t-1}\beta_1' + W. \tag{4}$$

Since $x_{t|t-1}$ summarizes the beliefs about x prior to observing y, it is a prior expectation in the Bayesian sense.

The next piece of the filter is to make use of a measurement equation which is the equivalent of conducting an experiment. The yield-per-unit effort is predicted using $\hat{x}_{t|t-1}$ and is compared to observed yield-per-unit effort. The yield-per-unit effort in period t is just y_t. At $t - 1$, one's beliefs about y_t are a consequence of equation (2),[3]

$$y_{t|t-1} = kx_{t|t-1} + v_y. \tag{5}$$

The point forecast of y_t is just the mean of (5),

$$\hat{y}_{t|t-1} = k\hat{x}_{t|t-1}. \tag{6}$$

From (5) and (6), the error in predicting y is the normal random variate,

$$e_{t|t-1} = y_t - \hat{y}_{t|t-1} = k\left(x_{t|t-1} - \hat{x}_{t|t-1}\right) + v_y, \tag{7}$$

where $\mathbf{E}[e_{t|t-1}] = 0$. The variance of the prediction error in catch-per-unit effort is $V_y + kR_tk'$, where V_y is the variance in the error of the yield

---

[2]   The quantities x and y could equally be interpreted as vectors. Then k is a matrix with the number of columns equal to the dimension of x and the number of rows equal to the dimension of y. Its transpose is k'. Similarly, the V's, P's, etc., are conformable matrices.

[3]   One could do this equally with equation (2r).

equation. One other statistic that will be needed is the covariance of $e_{t|t-1}$ and $x_{t|t-1}$.

$$\text{cov}\left(x_{t|t-1}, e_{t|t-1}\right) = \mathbf{E}\left[\left(e_{t|t-1} - 0\right)\left(x_{t|t-1} - \hat{x}_{t|t-1}\right)\right] = k'R_t, \quad (8)$$

where $\mathbf{E}[\ ]$ is the expectation operator. The variables $x_{t|t-1}$ and $e_{t|t-1}$ are correlated normal random variates, so they are jointly normal with the following distribution:

$$\begin{pmatrix} e_{t|t-1} \\ x_{t|t-1} \end{pmatrix} = N\left(\begin{pmatrix} 0 \\ \hat{x}_{t|t-1} \end{pmatrix}, \begin{pmatrix} V_y + kR_tk' & R_tk' \\ kR_t & R_t \end{pmatrix}\right). \quad (9)$$

Since observing the error in predicting catch-per-unit effort is the same as observing catch-per-unit effort, the posterior distribution of x (which is $x_{t|t}$) is just the same as $(x_{t|t-1}|e_t)$ or the conditional distribution of $x_{t|t-1}$, given the observed e_t. To summarize, equation (9) gives the joint distribution of the normal variates, $e_{t|t-1}$ and $x_{t|t-1}$. We seek the conditional distribution of $x_{t|t-1}$, given $e_{t|t-1}$. The conditional distribution for a joint normal can be found in a standard text or the article by Meinhold and Singpurwalla (1983). From the formula for the conditional distribution, $x_{t|t}$ is normally distributed with mean and variance given by

$$\begin{aligned} \hat{x}_{t|t} &= b_0 + b_1\left(\hat{x}_{t-1|t-1} - h_{t-1}\right) + R_tk'(V_y + kR_tk')^{-1}e_t, \\ P_{t|t} &= R_t - R_tk'(V_y + kR_tk')^{-1}kR_t. \end{aligned} \quad (10)$$

Equation (10) gives the mean of $x_{t|t}$ as a function of e_t — the observed observation error. Equation (10) is the Kalman filter for this simple fishery model. To recapitulate: For any set of parameters, β, δ, k, etc., and any $x_{0|0}$ and $P_{0|0}$, one can calculate all $x_{t|t}$'s by use of equation(10). That calculation is simply the algebra of conditional expectations and nothing more.

One estimates equations (1) to (3) by the maximum likelihood method. Equation (3) can be estimated by ordinary least squares, quite apart from the others. Unless one assumes contemporaneous correlation of the errors or the inclusion of endogenous variables (such as current price), as we will do later, the Gauss-Markov theorem assures that ordinary least squares is best. The other two equations are estimated by Kalman filter/maximum likelihood.

The maximum likelihood method requires the choice of parameters (β, k, x_0, P_0) to maximize the likelihood of what one observes. The only observable variable is y_t, and the likelihood of the sample is just the product of the likelihoods of the T observations.

A typical observation consists of y_t and $\hat{x}_{t|t-1}$. The likelihood of y is the same as $kx_{t|t-1} + v_y$ which is a normal variate with mean $k\hat{x}_{t|t-1}$ and variance $V_y + kR_tk'$. Letting u_t be the observed residual $u_t = y_t - k\hat{x}_{t|t-1}$, the likelihood of the t-th observation is

$$L_t = \frac{1}{\sqrt{2\pi(kR_tk' + V_y)}} e^{-u_t\left(V_y + kR_tk'\right)^{-1}u_t}. \tag{11}$$

The likelihood of the sample is just $L(x_{0|0}, P_{0|0}, \delta, \beta, k, V_y, W) = \prod_{t=1,T} L_t$ which is the likelihood of observing all T observations of y. In practice, for any set of parameters, one evaluates L by first using the filter, equation (10), to find all of the x's and P's. The x's and P's are used to calculate R_t and u_t. Finally, the L_t's are calculated and multiplied together from $t = 1, \ldots, T$ (or the logs of L_t are summed) to get the likelihood function. The maximum likelihood method is to use numerical methods to find the values of the parameters that maximize the likelihood function. We were not able to find algebraic expressions for the derivatives of this function with respect to the parameters, so we have no proof that the function is concave or that convergence is guaranteed.

The Kalman filter can be easily extended to use sampling information as well as catch-per-unit information. Halibut stocks are predicted by an age-cohort analysis. Let γ_t be the International Pacific Halibut Commission (IPHC) stock estimates made by this method alone, and assume that v_γ is their error with variance V_γ. Thus,

$$\gamma_t = x_t + v_\gamma \tag{12}$$

is a second measurement equation. The mechanics of the filter are as before except that V is now a matrix of v_y and v_γ, and u_t is a vector. One could enlarge this model to the belief that γ is a linear function of the true stock without undue computational burden. This combination of the filter and the sampling information gives the minimum mean square error way to use both cohort and catch-per-unit effort data. The actual series we have from IPHC already combines both of these methods in a nonoptimal fashion, so we do not pursue this any further.

In contrast to these methods, the methods in the literature are to either ignore the dynamic and stochastic nature of the fish stock (e.g., Bell, 1972) or make a clever substitution to eliminate the fish stock and ignore its

stochastic nature (e.g., Spence, 1973). There is little to be said in favor
of assuming fish stocks to be in equilibrium over a long sample period,
so there is little to be said for estimations of yield-effort curves based on
that assumption. The problem with Spence's ingenious method is that he
suppresses the error in the measurement equation.[4] On the other hand,
dispensing with an error term before one makes felicitous substitutions is
certainly well within the reduced form tradition of econometrics (add your
error terms when it is convenient). Short of the filter techniques proposed
here, Spence's method is certainly the next best.

In the next section we apply this Kalman filter/maximum likelihood
algorithm to a more realistic model of the Pacific Halibut fishery.

Application to the Pacific Halibut fishery

The IPHC was established in 1923 by a treaty between Canada and the
United States to rehabilitate and maintain Pacific Halibut stocks at or
near maximum sustainable yield. The fishery consists of four separately
managed areas. Since the IPHC cannot directly observe the stock of
halibut, it relies on changes in catch-per-unit effort and age composition
studies to manage the resource. The management tools used by IPHC are
gear restrictions, size limits, the regulation of incidental catch (IC), and an
annual quota on total catch. Although halibut are exploited by a variety
of vessel types that are shared with other fisheries, only one type of gear
— longline skate (a setline) — has been in use since the early days of the
fishery. The biology of the fishery is such that fishermen exploit a large
number of year classes simultaneously. For this reason, Crutchfield (1981)
states that the halibut fishery is "ideally characterized by the traditional
biomass-fishery model".

The model employs annual data on the halibut fishery from 1936 through
1982 for the two most important areas numbers 2 and 3. Table 1 gives the

---

[4] Spence writes his state equation as $x_{t+1} = F(x_t) - c_t$, where F is natural growth. In filter
terms, an observation equation would be Spence's catch equation, $c_t = F(x_t)[1 - e^{-\lambda E_t}]$,
with catch as a function of fishing effort and a parameter, λ. Let $z_t = c_t[1 - e^{-\lambda E_t}]^{-1}$;
then $z = F$ and one estimates $z_{t+1} = F[z_t e^{-\lambda E_t}] + w$. This appears to successfully
eliminate the unobservable from the equation. Spence reaches this simple result by writing
$z = F$ rather than $z = F + v$; that is, he ignores the error in measuring stock by catch
per (a function of) unit effort. Including this error in the measurement equation leads to
a much more complicated result. Put differently, one does not know stock. It is a random
variable derived from a stochastic process, and z provides only an estimate of it. If one
carries through the algebra, $z_{t+1} = F[(z_t - v_t)e^{-\lambda E_t}] + w_{t+1} + v_{t+1}$. Even if one does
not bother much with the problems caused by nonlinear F, this is still an autocorrelated,
lagged dependent variable problem in need of some attention.

Table 1: Pacific Halibut Fishery Model.

BIOMASS

$$\text{Ln}(\text{biomass}_{2,t}) = s_1 + s_2 \cdot \text{Ln}(\text{biomass}_{2,t-1} - \text{catch}_{t-1} - IC_{2,t-1}) + w_t.$$

CATCH

$$\text{Ln}(\text{catch}_2/\text{day}_{2,t}) = c_1 + c_2 \cdot \text{Ln}(\text{biomass}_{2,t})$$
$$+ c_3 \cdot \text{Ln}(\text{effort}_2/\text{day}_{2,t}) + v_{ct}.$$

EFFORT

$$\text{Ln}(\text{effort}_2/\text{day}_{2,t}) = \text{ef}_1 + \text{ef}_2 \cdot \text{Ln}(\text{biomass}_{2,t})$$
$$+ \text{ef}_3 \cdot \text{Ln}(\text{halprice}_t)$$
$$+ \text{ef}_4 \cdot \text{Ln}(\text{sablepr}_t)$$
$$+ \text{ef}_5 \cdot \text{Ln}(\text{salmonpr}_t) + v_{et}.$$

HALIBUT PRICE

$$\text{Ln}(\text{halprice}_t) = h_1 + h_2 \cdot \text{Ln}(\text{catch}_{2,t} + \text{catch}_{3,t} + \text{catch}_{4,t})$$
$$+ h_3 \cdot \text{Ln}(\text{pincome}_t) + h_4 \cdot \text{Ln}(\text{holdings}_t) + v_{ht}.$$

Notes: Subscripts indicate area and year.
IC is incidental catch.

equations of the complete model which is a good deal more complicated than the simplified model of section 2. Table 2 gives the definitions of all the variables.

The first equation is the equation for the evolution of the unobservable fish biomass (thousand metric tons, t.m.t.) Unlike the simple model, the state equation is not linear in the state variable, biomass. The functional form chosen for the stock equation allows slower growth when the biomass is larger (which the linear form does not), but it still does not capture the backward bending part of the growth curve or allow for a maximum biomass. Catch is what is caught by boats trying to catch halibut, while incidental catch, taken as exogenous, is what is caught by boats targeting other species. The stock equation is a feasible improvement over the linear form.

The nonlinearity of the biomass equation leads to filters that are based on the probability density function of biomass. This will likely result in nonlinear filters that are burdensome to implement. To maintain compu-

Table 2: Pacific Halibut Data and Source.

Variable	Description	Source
Catch	Quantity of halibut caught by the longline fleet in management areas 2, 3, and 4. Expressed in round weight metric tons.	Myhre, 1977; IPHC Annual Reports, 1977 to 1982.
Incidental catch	Quantity caught by other fishers in management areas 2, 3, and 4.	Myhre, 1977; IPHC Annual Reports, 1977 to 1982.
Effort	Number of longline skates used to catch halibut in management areas 2 and 3. Expressed in 100 skates.	Myhre, 1977; IPHC Annual Reports, 1977 to 1982.
Season length	Number of days between the opening and closing of the season for management areas 2 and 3.	Skud, 1977; IPHC Annual Reports, 1977 to 1982.
Halibut price	Average exvessel price of Pacific Halibut for areas 2, 3, and 4. Deflated using the Implicit Price Deflator for GNP (IPD) with base year = 1972.	IPHC Annual Report, 1982; U.S. President, 1970 and 1984.
Salmon price	Average exvessel price of salmon, all species, North Pacific Ocean. Deflated using IPD with base year = 1972.	United States Department of Commerce, 1975; Orth, et al., 1981; Fisheries of the United States, 1976 to 1982.
Sablefish price	Average exvessel price of sable-fish, North Pacific Ocean. Deflated using IPD with base year = 1972.	NMFS-Fishery Statistics of the United States, 1939 to 1956; U.S. Department of Commerce, 1974.
Cold storage holdings	Beginning of season holding of frozen Pacific Halibut expressed in round weight metric tons.	NMFS-Fishery Statistics of the United States; NMFS-Fishery Industries of the United States
Per capita income	U.S. Per Capita personal disposable income in 1972 dollars.	U.S. President, 1969, 1980, 1984.
Halibut biomass	IPHC estimates of Pacific Halibut Biomass from cohort and catch-age analysis, management areas 2 and 3.	Deriso and Quinn, 1983; Hoag and McNaughton, 1978; Quinn, et al., 1985.

tational simplicity, the "extended" Kalman filter is derived from the linearized state equation. It maintains filter linearity in the state (biomass) variable. First, we treat the state variable as the natural log of biomass. Its equation is linearized through a first-order Taylor series expansion about the mean of biomass, $\hat{x}_{t-1|t-1}$.

The size of the error introduced by linearizing the logged biomass equation depends on the size of catch relative to biomass. As catch increases relative to biomass, the error increases; and this is compounded by the degree of nonlinearity of biomass in the true biomass growth equation. In the case of the Pacific Halibut fishery, direct and incidental catch is a small proportion of the biomass (roughly 15 percent), so the error in estimating biomass via the extended Kalman filter is expected to be small. Although other methods of dealing with nonlinear state equations exist, they are more complicated and may not reduce error to the extent that would justify the additional effort needed to implement them. According to Gelb (1974), the extended Kalman filter "has been found to yield accurate estimates in a number of important practical applications".

The catch-per-day equation is just the classic equation of yield-effort fishery economics generalized to permit arbitrary, but constant, elasticities of catch per day with respect to stock and effort per day.

Effort per day in the halibut fishery is modeled as dependent upon current market forces. Current biomass and exvessel price directly affect current profitability, and sablefish and salmon prices give the opportunity cost of fishing. The gear used for fishing halibut (skate soaks) is not so specialized as to preclude the same vessels switching from one fishery to another. It is believed that this intraseason switching of target species is the main form of exit and entry in this fishery. Thus, current variables are expected to have a large effect on effort per day.

In all, there are six measurement equations — three each for management areas 2 and 3 of the Pacific Halibut fishery. Each management area also has a biomass equation. There is also an exvessel price equation to represent the demand for halibut at dockside. The parameters of these equations are estimated using the full information-Kalman filter/maximum likelihood technique. The technique of the previous section is generalized to include two state variables (log stock in each area), an extended filter, seven endogenous variables, and six measurement equations. The likelihood for an observation is just as in full information/maximum likelihood method except that the term kR_tk' (the variance induced by the uncertainty about the stock) is added to the usual variance covariance matrix.

The estimates of stock biomass are constructed from the logged biomass estimates and their variances, which are obtained from the Kalman filter. Since the conditional and updated estimates of the log of stock biomass in each time period are distributed normal with mean ν_t and variance P_t, the conditional and updated estimates of stock in each time period, x_t, are distributed log-normal with a mean of

$$\mathbf{E}(x_t) = e^{\nu_t + (1/2)P_t} \tag{13}$$

and a variance of

$$V(x_t) = e^{2\nu_t + P_t} \left(e^{P_t} - 1 \right). \tag{14}$$

Results

The parameter estimates of the model obtained from the numerical optimization using the Kalman filter and the Davidon-Fletcher-Powell optimization method are presented in Table 3. The standard errors were computed by means of a bootstrap,[5] but the histograms of the replicated parameter estimates tell a somewhat different story. The histograms do not appear to be symmetric and generally indicate that parameter values are very much less likely to be zero than might be concluded from the t ratios.

A parameter estimate of one associated with the exponent on the biomass equation is the demarcation between higher stock giving higher and lower growth. One cannot reject the simple model of linear growth for area 2, though one can reject it for area 3. The catch-per-day equation in area 2 is nearly linear in biomass and in effort per day (and they are not significantly different from unity in area 3) so, again, the very simple model seems acceptable. Effort per day is increasing in biomass and halibut price, which is as one would expect, but the effects of competing species are of uncertain sign (and this is true in both areas). Finally, the demand curve slopes down and, when one examines the bootstrap replicates, significantly so. Holdings depress price while per-capita income increases it, which is as it should be. Neither of these latter two variables are significantly different from zero. In summary, the parameter estimates are about what one should expect and are much closer to the estimates from a naive model than one would have first thought likely.

---

[5] The 92 percent, bias corrected, centered confidence intervals were also computed this way, using the method suggested by Efron (1982).

Table 3: Parameter Estimates of the Pacific Halibut Fisher Model.

Parameter	Estimate	Standard deviation	92 percent C.I.[a]	
Area 2 biomass				
Intercept — s_1	0.304	0.242	0.032	0.309
Escapement 2 — s_2	0.956	0.050	0.931	1.08
Area 3 biomass				
Intercept — s_1	1.43	0.726	1.40	5.05
Escapement 3 — s_2	0.727	0.100	0.278	0.735
Area 2 catch per day				
Intercept — c_1	−4.15	20.35	−86.0	−0.397
Biomass 2 — c_2	1.03	3.75	0.364	2.14
Effort 2 per day — c_3	0.876	0.428	0.008	0.981
Area 3 catch per day				
Intercept — c_1	−0.049	3.66	−6.42	2.16
Biomass 3 — c_2	0.643	0.893	−0.046	1.06
Effort 3 per day — c_3	0.751	0.619	0.172	2.14
Area 2 effort per day				
Intercept — ef_1	−17.41	20.7	−82.8	−2.07
Halibut price — ef_2	1.47	0.779	0.736	3.11
Biomass 2 – ef_3	2.78	3.60	0.968	14.8
Sablefish price — ef_4	0.083	0.426	−0.388	1.09
Salmon price — ef_5	−0.981	0.516	−0.980	1.14
Area 3 effort per day				
Intercept — ef_1	−7.30	6.00	−25.1	−6.0
Halibut price — ef_2	1.08	1.34	0.614	6.64
Biomass 3 — ef_3	1.23	0.598	0.874	2.84
Sablefish price — ef_4	−0.536	0.291	−0.832	−0.379
Salmon price — ef_5	0.156	0.892	−0.930	2.57
Exvessel halibut price				
Intercept — h_1	12.8	12.8	9.17	74.4
Total catch — h_2	−0.831	0.839	−5.45	−0.650
Income — h_3	0.430	1.04	−1.08	0.839
Holdings — h_4	−0.124	0.131	−0.187	0.375

[a]Bias corrected 92 percent central confidence interval about the point estimate.

The biomass estimates

Estimates of Pacific Halibut biomass (Table 4) for management area 2 and management area 3 are a byproduct of the Kalman filter/maximum likelihood estimation. This subsection presents the estimates from the Kalman filter. These estimates are compared to biomass estimates obtained from a recent IPHC publication (Quinn, Deriso, and Hoag, 1985). The IPHC biomass estimates were derived from catch-age analysis.

Figure 1 is a plot of the updated biomass estimates for area 2 against time in years. Estimates of area 2 of biomass from Quinn, Deriso, and Hoag are also plotted on this graph. The pattern of biomass estimates from the Kalman filter methodology strongly resembles the pattern of biomass estimates given by IPHC. According to IPHC estimates, the peak biomass occurs in 1955. The Kalman filter biomass estimates increase dramatically from 1943 to 1954 and decrease sharply from 1955 to 1960. The IPHC biomass estimates begin to rise again in 1980 while the Kalman filter biomass estimates begin to rise in 1977.

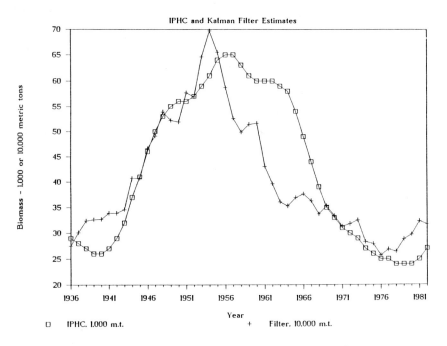

Figure 1: Halibut Biomass in Area 2.

For area 2, the regression of IPHC estimates (IPHC2) on the Kalman filter estimates (BIO2) resulted in the following relationship (standard deviation in parentheses):

$$BIO2_t = 110 + 6.9 \, IPHC2_t \,,$$
$$(24) \quad (.54) \tag{15}$$
$$R^2 = .78; \quad n = 47; \text{ residual standard error} = 55.3.$$

Estimation of the model parameters and variance matrix using the Kalman filter/maximum likelihood technique resulted in biomass estimates that are 10 times larger than those of the IPHC. This discrepancy becomes apparent immediately in the Kalman filter recursions by examining the conditional and updated biomass estimates and variances for 1936. Given the bioeconomic model, the biomass estimates have very low variances. As a result, the IPHC estimates do not fall in the confidence intervals of the Kalman filter biomass estimates.

Table 4, column 3, presents the updated estimates of area 3 biomass. Figure 2 presents a plot of the Kalman filter biomass estimates and the IPHC estimates (Quinn, Deriso, and Hoag, 1985) against time in years. Both patterns exhibit two peaks and two valleys. The IPHC biomass series peaks in 1946 and 1961 while the Kalman filter series peaks in 1952 and 1960. Both series declined beginning in 1962. The biomass begins to increase again in 1976 according to the IPHC series and in 1975 according to the Kalman filter series. The Kalman filter series predicts a much faster increase in biomass after 1979 than does the IPHC series.

The regression of IPHC biomass estimates (IPHC 3) on the Kalman filter biomass estimates (BIO 3) resulted in the following relationship (standard deviation in parentheses):

$$BIO3_t = 25 + .79 \, IPHC3_t \,,$$
$$(12) \quad (.18) \tag{16}$$
$$R^2 = .29; \quad n = 47; \text{ residual standard error} = 27.$$

The Kalman filter biomass estimates are approximately the same magnitude as the IPHC estimates. While the explanatory power of equation (16) is not as great as in equation(15), the estimates of biomass in area 3 after 1975, predicted by the Kalman filter model, agree with the increases in biomass after 1975 that were noted throughout the fishery in 1980 through 1986 (van Amerongen, 1985; *Alaska Fisherman's Journal*, 1984 and 1986).

The biomass estimates for both areas follow the pattern of IPHC biomass estimates. Area 2 biomass estimates from the Kalman filter are about 10

Table 4: Pacific Halibut Biomass Estimates in Area 2 and Area 3, 1936–1982 (in round weight per 1,000 metric tons).

Year	Biomass (t/t) area 2	Biomass (t/t) area 3
1936	274.83	47.82
1937	301.19	51.07
1938	323.68	53.11
1939	325.85	54.96
1940	327.39	56.86
1941	338.91	65.11
1942	338.63	62.09
1943	345.65	64.56
1944	407.12	58.00
1945	406.73	67.30
1946	465.85	81.84
1947	490.99	82.41
1948	539.13	102.56
1949	521.47	95.14
1950	519.19	95.66
1951	576.91	113.75
1952	569.22	127.99
1953	646.30	121.78
1954	697.27	116.97
1955	655.42	88.57
1956	587.30	78.40
1957	525.63	69.68
1958	499.34	78.29
1959	513.94	103.89
1960	516.30	112.72
1961	431.01	100.47
1962	396.33	97.43
1963	361.30	103.45
1964	352.85	88.74
1965	368.65	73.36
1966	375.76	88.10
1967	363.38	62.39
1968	337.12	52.07
1969	352.72	56.77
1970	334.27	55.42
1971	313.45	43.96
1972	317.82	40.63
1973	325.18	29.42
1974	282.88	23.37
1975	278.89	26.37
1976	257.39	28.40
1977	268.01	38.26
1978	263.87	43.07
1979	287.65	53.69
1980	296.53	98.89
1981	323.30	131.54
1982	316.73	175.82

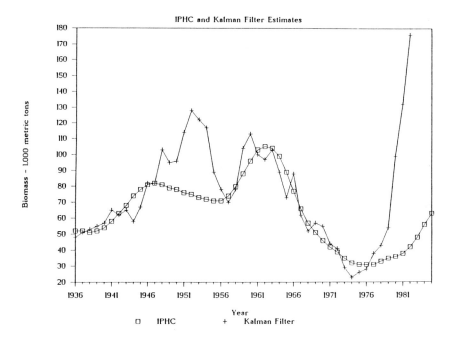

Figure 2: Halibut Biomass in Area 3.

times larger than the IPHC estimates although both series follow the same trend over time. In area 3, the IPHC and the Kalman Filter biomass estimates have the same magnitude. The time series of both estimates show two peaks, two valleys, and a sharp increase in biomass from 1980 to the present. For area 3, the filter estimates are much closer to the post-1975 fishing experience than are the IPHC estimates.

Maximum sustainable yield

The IPHC is charged with developing and maintaining halibut stocks at a level which provides maximum sustainable yield (MSY) to the fishery (Bell, 1978). The IPHC uses estimates of MSY and annual surplus production (ASP) to set annual quotas for management areas 2, 3, and 4. MSY is the maximum harvest that can be caught on a sustained basis without depleting the population. To achieve MSY, the quotas are set equal to 75 percent of ASP. ASP is the maximum potential change in biomass from the previous year to the current year. ASP in the current year is equivalent to the catch in the previous year plus the change in biomass from the previous year to the current year. When catch is held below (above) ASP, then the biomass increases (decreases). This method is appropriate when stocks are below the level necessary to achieve MSY. To ensure that the quotas are below ASP, the ASP is multiplied by 0.75.

The ASP, MSY, Maximum Sustainable Catch (MSC), and the catch that maximizes the present value of revenues to the fishery were calculated using the estimated model. The results for each area along with the actual quotas set by IPHC are presented in Table 5 for the years 1984, 1985, and 1986. The area 2 biomass is below MSY. However, it is close to the level that provides the highest revenues. The 75 percent ASP and the revenue maximizing catch in area 2 are approximately equal and substantially greater than the actual quota recommended by IPHC. The area 3 biomass is above MSY. The allowable catch recommended by the 75 percent ASP and the revenue maximization of area 3 provides for a reduction in biomass to levels at and below MSY, respectively. Both catch levels are significantly above the actual quota.

Revenues in both areas can be increased by raising the quotas without compromising the productivity of the fishery. In both management areas, a policy that restricts catch to achieve MSY will not benefit halibut fishers. This result explains the events of the mid-1980s when unprecedented increases in catch-per-unit effort resulted in record volumes of halibut caught and a severe depression in halibut prices and income. The current problem in the Pacific Halibut fishery is the accurate forecasting and control of incidental catch of halibut. The size of the incidental catch has a large impact on the biomass and the recommended quota. The degree to which IPHC regulations will benefit fishers in the future will depend on their ability to recommend quotas consistent with maximum revenue, or income, and their ability to estimate and control the amount of incidental catch.

Conclusions and extensions

Filter methods were shown to provide a reasonable basis for estimation of fishery models. They are equally useful for other economic cases of unobserved stocks.

Extension: Other examples

Other examples of important stochastic stock problems in resources and agriculture include aliens, crop acreage, and undiscovered reserves of an

Table 5: Comparison of Recommended and Actual Quotas (1,000 metric tons).

Year/ Area	75 percent ASP	Maximum revenue	Actual quota	Biomass under actual quota
1984				
Area 2	20.4	19.4	9.1	352.7
Area 3	39.4	52.1	13.7	256.9
1985				
Area 2	20.7	20.1	11.8	368.8
Area 3	36.4	35.2	17.5	291.3
1986				
Area 2	21.0	19.9	13.9	382.6
Area 3	33.8	26.2	21.0	318.0
Area	Maximum sustainable yield			
Area 2	732.3			
Area 3	211.5			
	Maximum sustainable catch			
Area 2	32.2			
Area 3	57.7			

exhaustible resource. The method could also be used to estimate aggregate capital. Each of these models is sketched in turn.

To make an estimate of resources which remain to be found in some areas, a planner could reason as follows. The prior is $(\hat{x}_0, P_0)$ — the quantity of the resource to be discovered and its variance. It might just as well be a vector of the types of resources and their covariances. Again, h_t is found with exploration effort, E_t, so the observation equation is $h_t = f(x_t, E_t)$ and the state equation is simply $x_{t+1} = x_t - h_t$. The measurement equation has the same justification as the fishing measurement equation: It is easier to find an exhaustible resource when there is more of it to be found. The state equation is the exhaustible resource state equation. Assume that the measurement equation is just the familiar $h/E = kx$. Substitute the definitions of $\hat{x}_{t|t-1}$ and e_t into equation (10), the filter; and subtract h_t from both sides to get

$$\hat{x}_{t|t} - h_t = \hat{x}_{t|t-1} + R_t k (V + k R_t k')^{-1} (y_t - k \hat{x}_{t|t-1}) - h_t. \qquad (17)$$

Equation (17) is the estimate of the mean stock remaining to be discovered after the findings of period t, that is, $\hat{x}_{t+1|t}$. Since $y_t = h_t/E_t$, $(\partial\hat{x}_{t|t+1})/(\partial h_t) > 0$ whenever

$$R_t k'(V + kR_t k')^{-1}E_t^{-1} > 1. \tag{18}$$

Looking at (9), this can be re-expressed as (19)

$$\frac{\text{cov}\left(e_{t|t-1}, x_{t|t-1}\right)}{\text{var}\left(e_{t|t-1}\right)} > E. \tag{19}$$

Increasing discoveries increases one's estimate of stock when surprises in discovery per unit effort are highly correlated with stock and when the absolute value of effort or the variance of stock is low.

The multivariate expansion of this model could include good and bad grades of the resource. The discovery equation would then have two k's, one for each grade. Discovery effort would result in both good and bad grades being discovered, in proportion to their difficulty to discover, k, and their abundance, x. This gives a model without the usual odious assumption that good grades are discovered first.

Undocumented crop acreage is a real problem in California. Marijuana is often alleged to be an important (sometimes the important) cash crop of the northern timber growing regions. The natural agricultural model is a stock adjustment model. Unobservable production, x, is a function of past production, observable price, and observable apprehension expenditure. The latter variable represents a very severe cost to the grower. In addition to expenditure on enforcement, one also observes enforcement success — tons seized. Let y be tons seized per dollar of enforcement effort. As before, $y = kx$ is the observation equation. There are many problems with this example — particularly rapid technical progress in crop cultivation and detection avoidance and equally rapid progress in detection through aerial surveillance.

Undocumented aliens are definitely not directly observable. Torok and Huffman (1986) examined U.S.-Mexican trade in winter vegetables and undocumented immigration. Their model shows that the same work force picks tomatoes in both countries so that the United States will import either labor or tomatoes. To estimate their structural model, they substitute apprehensions and apprehension effort for the actual unobservable stock of laborers and supply of labor. Their model could be cast in the filter mode by adding a state equation and treating the stock of those undocumented in the other equations. A plausible state equation would be: The stock of Mexican agricultural laborers (undocumented) is determined by the ·past stock and wage and unemployment differentials between the United States and Mexico. Undocumented immigration is a complicated matter. Both the actual number of aliens who do not return home and the size of the network that safely imports them contribute to what would be measured as stock. The observation equations are the demand and supply of undocumented labor. The demand equation in their model (simplified) is $\ln(A) = c_0 + c_1 \ln(P) + c_2 \ln(w)$, where A is apprehensions, P is tomato price, and w is the wage rate. The c's are constant. The model is derived from a simple yield effort model: $\ln(A) = k \ln(BP) + \ln(N)$, where BP is apprehension effort by the Immigration and Naturalization Service and N is the stock of undocumented laborers. (They also assume that labor supply is proportional to N, so the demand equation for apprehensions really is the labor demand equation.) The natural way to expand the model would be to avoid the substitution of apprehensions for labor quantity and write $\ln(N) + k_1 \ln(BP) = c_0 + c_1 \ln(P) + c_2 \ln(w)$ and $\ln(A) = k_2 \ln(BP) + \ln(N)$ as the two observation equations. The two different k's allow for the difference between labor supply and the stock of undocumented laborers. A third observation equation would be the labor supply equation which can be handled in a similar fashion.

The last example is the capital stock. The state equation is $K_t = K_{t-1} - \delta K_{t-1} + I_t$ or capital is depreciated at the unknown rate, δ, and replenished by investment, I. The usual method for constructing such a sequence is to take K_0 as some reasonable estimate and infer δ from depreciation data for various industries. Although the resultant numbers are used as the capital stock, they are clearly estimates with substantial probable error. Viewed this way, $Y = F(K, L, M)$; the aggregate production function in terms of capital (labor and materials) is just a measurement equation for the capital stock. The filter will produce stock estimates, estimates of the production function, and estimates of the reliability of the stock estimates.

Conclusions

Unobservable stochastic variables play a major role in many applied economic fields. In fishery economics, the major explanatory variable is the unobserved fish stock. Sampling methods based upon population dynamics are almost always supplemented with information inferred from the economic activity of the harvesting agents. In this paper, we used just the information from economic activity to infer stock. Our Pacific Halibut biomass estimates generally agree in pattern, though not magnitude, with the estimates made by IPHC. In the immediate postsample period, the filter estimates for area 3 provide more accurate predictions of halibut catch-per-unit effort than the IPHC estimates. The experience in the fishery was closer to our expectations than it was to those of the IPHC. We view this as providing some evidence for the utility of this method.

On the other hand, the vast disparity in our stock estimates and those of the IPHC suggests that our estimates ought to be called effective stock, that is, the unobserved variable that correlates well with fishing success.

This paper also provides several other examples of unobserved variables in economics. In each case the Kalman filter/maximum likelihood approach is a promising method for preserving the stochastic variability and endogeneity of the model during estimation. This method provides more information on the dynamics of the unobserved variable than has been available or used in past studies.

References

Alaska Fisherman's Journal (1984): *Extending the Halibut Season*, Vol. 7, No. 11. Seattle, Washington.

Alaska Fisherman's Journal (1986): *The Year of the Halibut*, Vol. 9, No. 1. Seattle, Washington.

Amerongen J. van (1985): "Good News/Bad News for Halibut Fleet: High Catches/Low Returns", *Alaska Fisherman's Journal* 8(3).

Bell F.H. (1978): "The Pacific Halibut: Biology, Fishery and Management", Technical Report No. 16, International Pacific Halibut Commission, Seattle, Washington.

Bell F.W. (1972): "Technological Externalities and Common Property Resources: An Empirical Study of the U.S. Northern Lobster Fishery", *Journal of Political Economy* 80: 148–158.

Berck P., and J.M. Perloff (1984): "An Open-Access Fishery with Rational Expectations", *Econometrica* 52(2): 489–506.

Crutchfield J.A. (1981): *The Public Regulation of Commercial Fisheries in Canada: The Pacific Halibut Fishery*, Economic Council of Canada, Case Study No. 2, Technical Report No. 17.

Deriso R.B., and T.J. Quinn II. (1983): "The Pacific Halibut Resource and Fishery in Regulatory Area 2. II. Estimates of Biomass, Surplus

Production and Reproductive Value", Scientific Report No. 67, International Pacific Halibut Commission, Seattle, Washington.

Efron B. (1982): "The Jackknife, the Bootstrap and Other Resampling Plans", CVMS-NSF Regional Conference Series in Applied Mathematics, Society of Industrial and Applied Mathematics, Philadelphia, Pennsylvania.

Fullenbaum R.F., E.W. Carlson, and F.W. Bell (1971): "Economics of Production from Natural Resources: Comment", *American Economic Review* 61(3): 483–491.

Gelb A. (1974): *Applied Optimal Estimation*, Analytic Sciences Foundation, Cambridge: Massachusetts Institute of Technology Press.

Harvey A.C. (1981): *Time Series Models*, New York: Halsted Press.

Hoag S.H., and R.J. McNaughton (1978): "Abundance and Fishing Mortality of Pacific Halibut, Cohort Analysis, 1935–1976", Scientific Report No. 65, International Pacific Halibut Commission, Seattle, Washington.

Meinhold R.J., and N.D. Singpurwalla (1983): "Understanding the Kalman Filter", *The American Statistician* 37(2): 123–127.

Myhre R.J. (1977): "The Pacific Halibut Fishery: Catch, Effort and CPUE 1929–1975", Technical Report No. 14, International Pacific Halibut Commission, Seattle, Washington.

Orth F.L., J.R. Wilson, J.A. Richardson, and S.M. Piddle (1981): "Market Strucrure of the Alaskan Seafood Processing Industry", Vol. 2, Finfish, Sea Grant Report 78–14, University of Alaska, Fairbanks, Alaska, March.

Quinn T.J. II, R.B. Deriso, and S.H. Hoag (1985): "Methods of Population Assessment of Pacific Halibut", Scientific Report No. 72, International Pacific Halibut Commission, Seattle, Washington.

Skud B.E. (1977): "Regulation of the Pacific Halibut Fishery 1924–1976", Technical Report No. 15, International Pacific Halibut Commission, Seattle, Washington.

Smith V.L. (1986): "Economics of Production From Natural Resources", *American Economic Review* 58: 409–431.

Spence A.M. (1973): "Blue Whales and Applied Control Theory", Technical Report 108, Institute for Mathematical Studies in the Social Sciences, Stanford, California, 1–35.

Torok S.J., and W.E. Huffman (1986): "U.S.-Mexican Trade in Winter Vegetables and Illegal Immigration", *American Journal of Agricultural Economics* 6(2): 247–260.

Stochastic Models and Option Values
D. Lund and B. Øksendal (Editors)
© Elsevier Science Publishers B.V. (North-Holland), 1991

Optimal Decisions With Reduction of Uncertainty over Time — An Application to Oil Production

Gunnar Stensland[1] and Dag B. Tjøstheim[2]

[1]*Chr. Michelsen Institute, N-5036 Fantoft, Norway*
[2]*University of Bergen, N-5014 Bergen, Norway*

1. Introduction

Sequential evaluation and decision problems must frequently be solved under uncertainty. If there is a gradual reduction of uncertainty over time, it is important to take this into account. In the present paper we suggest a general procedure for doing that. Our efforts are motivated in particular by problems occurring in oil production with uncertain production profiles. However, the method remains valid for general decision problems where the uncertainty is resolved in time.

A typical (idealized) production profile is shown in Figure 1a. At the beginning there is startup phase when the first production wells are drilled. Subsequently follows a plateau phase in which production reaches its maximum level. During this period the pressure in the reservoir decreases. When it becomes lower than a certain threshold, it is insufficient to support maximum production, and the production starts to decline. The field is abandoned when, roughly speaking, production costs are larger than the value of production. Typically the decision of abandonment has to be taken in the face of uncertain future production quantities and uncertain oil prices.

When exploiting a resource such as petroleum, information about the size of the resource and the productivity is collected continuously. This means that the production profile of the field has to be revised as new information concerning the reservoir is gathered. It is not uncommon that a very substantial revision takes place. At time $t = 0$, when production starts, this information is not available. However, the uncertainty may be expressed in terms of a number of possible production scenarios as in Figure 1b. The relative likelihoods for the scenarios are determined by geophysical/geological knowledge prior to production. The scenarios

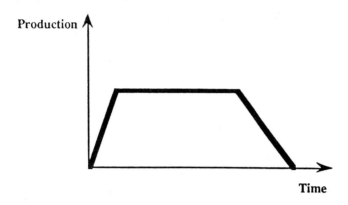

Figure 1a: Idealized production profile.

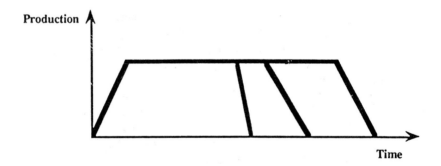

Figure 1b: Three production scenarios (all having the same plateau level).

should only be taken as a skeleton structure. In our model any intermediate production quantity can be obtained by weighting the scenarios.

As time progresses and the information increases, the probability distribution over the scenarios will be gradually narrowed, and when information is complete, we are left with one scenario as the true one. In Figure 2 is given a graphical plot of uncertainty resolution. It is taken from a publication of the Norwegian Oil Directorate (1987), and it covers both the time of production and the time prior to it. A problem with such plots is that it is not specified what "reservoir information" means or how it is measured. In this paper we will utilize information curves similar to that of Figure 2, but we will use a more concrete interpretation, in which an increase of information is specified in terms of percentage of variance explained.

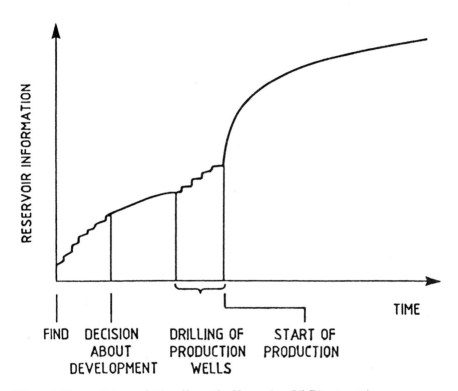

Figure 2: Uncertainty resolution (from the Norwegian Oil Directorate).

In spite of the uncertainty present in production problems most oil companies evaluate investment alternatives at $t = 0$ using standard deterministic net present value methods. Independent consultants such as Wood-MacKenzie do likewise. This is an open loop strategy, and the time of abandonment is determined at $t = 0$ as the point where the expected production profile meets the production cost. It results in a too late abandonment if one of the short production scenarios turns out to be true and too early if one of the long ones prevails. This means that there will be a systematic under-evaluation of the field. The downward bias will be particularly serious for marginal fields with production incomes just barely exceeding variable production costs. The effect will be considerably less for low cost fields and totally absent in the hypothetical case of a zero-cost field, where no decision is involved.

In practice of course such an open loop strategy is never followed. If the production takes an early downturn, it is stopped accordingly and vice versa. Our point is that this should be included also in the evaluation at time zero, which means that the production scenarios of Figure 1b and the information curve of Figure 2 (both available at time 0) should be taken into account. (In a private communication Peter Berck puts it this way: "One knows that one will know, but one does not know yet"). Actually, the main contribution of this paper is to combine the plots of Figures 1 and 2 in such a way that they can be utilized effectively in evaluating the economic potential of a field and in making optimal decisions.

Several authors have noticed the importance of uncertainty in producing from a resource. Olsen and Stensland (1988, 1989) use a geometric diffusion process with a negative trend to describe the production, and their model gives a closed form solution to some optimization problems with both the resource in place and the product price being uncertain. A disadvantage of this method is that it does not fully use the a priori information about uncertainty resolution. The more general aspect of this problem is sometimes referred to as "How to eat a cake with unknown size". See for example Kemp (1976), Loury (1978), and Hoel (1978). However, the learning mechanism presented in these papers is somewhat unrealistic.

In Deshmukh and Pliska (1980) and Gilbert (1979) exploration serves as a means of updating total reserves. In Aaheim (1986) and Pindyck (1980) the evolution of remaining reserves is described by the stochastic differential equation

$$dX_t = -U_t dt + dW_t \,, \tag{1.1}$$

where X_t is the remaining reserve, U_t is production, and W_t is a Wiener process. However, what we really need, is a model for the variation of production over time depending on remaining reserves and incorporating uncertainty resolution.

Most of the above papers are more concerned with macroeconomic problems than with revenue optimization for a single oil field. Related problems in microeconomics in general are discussed for example in Roberts and Weitzman (1981) with emphasis on optimal stopping problems in funding of

R&D projects. There are also several examples within the field of optimal search, where updating of information is done using Bayesian methods. See for example Rothschild (1974) and Burdett and Vishwanath (1988). Compared to these examples our problem differs on two points. First, we have more information about the speed of the uncertainty resolution at any stage, and second, a revenue at this stage is immediately revealed once the observations are available. In this paper, with the exception of Section 5, production from the field is the only uncertain variable. Hence there is no systematic risk present, and following the ideas behind the CAPM no risk premium is included.

Most of the above methods are based on continuous time stochastic differential equations where closed form solutions can be obtained in the simplest cases. We will use a discrete time approach akin to that of Stensland and Tjøstheim (1989). The discrete time formulation has the advantage that it leads to models which can be implemented on a computer, e.g. via dynamic programming. Moreover, the model is flexible, and a number of changes and extensions can be built into it. However, we do not by any means claim completeness of the model and method used. A number of interesting open problems and possible generalizations remain as indicated in Section 6.

2. Description of the method

As stated in the introduction, the two main components of our analysis are production profiles as in Figure 1 and resolution curves as in Figure 2. Our goal is to combine those two components in such a way that it is possible to evaluate oil fields and make optimal decisions under uncertainty. Typically the evaluation takes place before the production starts, but nevertheless one should take into account the uncertainty present in the problem and its reduction over time.

2.1 The system and its dynamics

To be more specific, assume that there are M production profiles associated with M production scenarios, and these have initial probabilities $\pi_1, \pi_2, \ldots, \pi_M$. Uncertainty resolution is described on a scale going from 0 to 1, where the extremes 0 and 1 correspond to none and complete resolution of uncertainty, respectively. We will denote by a_n the degree of uncertainty resolved at time n with $0 \leq a_n \leq 1$ and $a_0 = 0$. It will be

seen later that a_n can be interpreted in terms of percentage variance reduction of the scenario distribution as time develops. It should be realized
that most likely none of the given scenarios will give an exact description
of the production. The scenarios should be interpreted as a tool to generate the possible stochastic variation inherent in the problem. At each
instant of time the production will be computed as a weighted average of
the scenarios.

We will use a dynamic programming approach, and the state of our
system at each time point will be described by the location and shape of
the scenario distribution. The location of the scenario distribution will be
denoted by θ_n. For a symmetric distribution it represents the most likely
scenario and the center of the scenario distribution at time n. At time 0
it is reasonable to identify θ_0 with the scenario having the largest initial
probability, i.e. $\theta_0 = k$, where $\pi_k \geq \pi_j, j = 1, \ldots, M$. We denote by X_n
the true scenario at time n. Seen from time 0, X_n is a stochastic variable
which is distributed according to the narrowed scenario distribution located
at θ_n. More precisely it is described by

$$q_{ij}^{(n)} = \Pr(X_n = j|\theta_n = i), \quad i, j = 1, 2, \ldots, M. \tag{2.1}$$

Clearly, $q_{ij}^{(0)} = \Pr(X_0 = j|\theta_0 = i) = \pi_j$, the given initial distribution over
the scenarios. If the uncertainty is completely resolved at time n, then
$X_n = \theta_n$, and $q_{ij}^{(n)} = 1$ for $i = j$ and zero otherwise.

The core of the problem is now to devise a method for describing $\{\theta_n\}$
and $\{q_{ij}^{(n)}\}$ based on the uncertainty resolution curve generated by$\{a_n\}$.

The philosophy we have developed is perhaps easiest to explain if we have
a continuum of possibilities described by a probability density function over
the scenarios. This density function may be thought of as corresponding
to the continuous distribution of the resources at time zero. In fact the
production scenarios may be linked to the resource distribution via a simple
production model (e.g. Arps (1956)) or more complicated models depending
on geophysical variables. There are several possibilities for choosing the
continuous initial distribution, but here we will use the normal distribution.
This corresponds to a symmetric initial distribution, but continuous skew
densities can be incorporated as outlined in Appendix 1 of Bjørstad et al.
(1989). In the next subsection it will be demonstrated how a discrete and
symmetric initial distribution over the scenarios $1, \ldots, M$ can be converted
into a normal density. Moreover, we are currently investigating a more
direct method where it is possible to by-pass normality altogether.

To distinguish between the continuous and discrete case in the following the continuous versions of the basic variables θ and X will be starred. Assume that the initial normal distribution is given by $N(\mu_0, \sigma_0)$. We then take $\theta_0^* = \mu_0$ and X_0^* is distributed according to $q(X_0^*|\theta_0^*) = N(\theta_0^*, \sigma_0)$. It is not difficult to show that the results are independent of our choice of μ_0 and σ_0, and we may in fact (cf. Section 3.2) use $\mu_0 = 0$ and $\sigma_0 = 1$. This invariance property is reasonable since the normal distribution is used merely as a way of expressing the relative likelihood of the various scenarios. No special significance should be attached to its location and scale as the scenarios are not numerical quantities.

After production has started a new estimate $q(X_1^*|\theta_1^*)$ of the scenario distribution function at time 1 will be obtained. Again θ_1^* denotes the location (center) of the scenario distribution. It is reasonable to assume that $q(X_1^*|\theta_1^*)$ belongs to the same class of distributions as the initial one, i.e. $q(X_1^*|\theta_1^*) = N(\theta_1^*, \sigma_1)$ in our case. Seen from time 0, θ_1^* is a stochastic variable with its own distribution $p(\theta_1^*) = p(\theta_1^*|\theta_0^*)$. In the lack of other knowledge than the initial distribution it is reasonable to require that $E(\theta_1^*) = \theta_0^* = \mu_0$, and there remains the problem of choosing a distributional family for $p(\theta_1^*)$.

Within a Bayesian framework it is natural that p should be chosen as the conjugate distribution (cf. Raiffa and Schlaifer (1961) and DeGroot (1970)) to q. The class of conjugate distributions has the property that θ^* should be distributed according to the same family of distributions before and after more information has been collected. If $q(X^*|\cdot)$ is normal, the conjugate distribution $p(\theta^*)$ is also normal, when we assume the variance of to be known (DeGroot (1970), p. 167), so that in our case (for known variance) $p(\theta_1^*|\theta_0^*) = N(\theta_0^*, \beta_1)$, where $\beta_1^2 = \text{Var}(\theta_1^*)$. Analogously, if $q(X^*|\cdot)$ is a gamma distribution, the conjugate $p(\theta^*)$ is also a gamma distribution if we assume that one of the parameters in the distribution is known.

We will now show that we can in fact determine the variance β_1^2 of θ_1^* as well as the variance σ_1^2 in $q(X_1^*|\theta_1^*)$ from the uncertainty resolution curve $\{a_n\}$ if this is interpreted as percentage reduction of the variance in the scenario distribution $q(X_n^*|\theta_n^*)$. More precisely, for stage one the variance σ_1^2 is taken as

$$\sigma_1^2 = (1 - a_1)\sigma_0^2, \qquad (2.2)$$

so that $a_1 = (\sigma_0^2 - \sigma_1^2)/\sigma_0^2$. The total uncertainty as represented by the distributions q and p at stage 1 should be conserved when viewed from time

0, so that we require $\text{Var}(X_1^*) = \text{Var}(X_0^*) = \sigma_0^2$. However, the following identity holds for the conditional variables:

$$\begin{aligned}\text{Var}(X_1^*) &= \text{Var}(E(X_1^*|\theta_1^*)) + E(\text{Var}(X_1^*|\theta_1^*)) \\ &= \text{Var}(\theta_1^*) + E(\sigma_1^2) = \beta_1^2 + (1 - a_1)\sigma_0^2\,.\end{aligned} \quad (2.3)$$

Thus $\sigma_0^2 = \beta_1^2 + (1 - a_1)\sigma_0^2$, and it follows that $\text{Var}(\theta_1^*) = \beta_1^2 = a_1\sigma_0^2$. The distribution for θ_1^* is then completely determined.

Similarly we can generate distributions from step $n-1$ to n. Assume that at time n the scenario distribution is concentrated at θ_n^* which is distributed according to $p(\theta_n^*|\theta_{n-1}^*) = N(\theta_{n-1}^*, \beta_n)$ with $\beta_n^2 = \text{Var}(\theta_n^*|\theta_{n-1}^*)$. Again we interpret the numbers $\{a_n\}$ in terms of percentage reduction of the variance of the scenario distribution so that $\text{Var}(X_n^*|\theta_n^*) = \sigma_n^2 = (1 - a_n)\sigma_0^2$, or $a_n = (\sigma_0^2 - \sigma_n^2)/\sigma_0^2$. Requiring conservation of total uncertainty as seen from time 0 we have

$$\sigma_0^2 = \text{Var}(X_n^*) = \text{Var}(E(X_n^*|\theta_n^*)) + E(\text{Var}(X_n^*|\theta_n^*)) = \text{Var}(\theta_n^*) + \sigma_n^2\,, \quad (2.4)$$

and thus $\text{Var}(\theta_n^*) = a_n\sigma_0^2$. Conditioning on θ_{n-1}^* and applying the identity in (2.3) once more we have

$$\text{Var}(\theta_n^*) = \text{Var}(E(\theta_n^*|\theta_{n-1}^*)) + E(\text{Var}(\theta_n^*|\theta_{n-1}^*)) = \text{Var}(\theta_{n-1}^*) + \beta_n^2\,, \quad (2.5)$$

and hence $\beta_n^2 = \text{Var}(\theta_n^*|\theta_{n-1}^*) = (a_n - a_{n-1})\sigma_0^2$. Taking normality into account the n-step distributions are given by

$$\begin{aligned}q(X_n^*|\theta_n^*) &= N\left(\theta_n^*, \sqrt{1 - a_n}\sigma_0\right) \quad \text{and} \\ p(\theta_n^*|\theta_{n-1}^*) &= N\left(\theta_{n-1}^*, \sqrt{a_n - a_{n-1}}\sigma_0\right).\end{aligned} \quad (2.6)$$

If we look at the two extreme possibilities $a_n = 1$ and $a_n - a_{n-1} = 0$, we see that for full information resolution ($a_n = 1$) at step n, there will be no uncertainty left in the scenario distribution $q(X_n^*|\theta_n^*)$, whereas if no information is revealed ($a_n = a_{n-1}$), we have $\theta_n^* = \theta_{n-1}^*$ and $q(X_n^*|\theta_n^*) = q(X_{n-1}^*|\theta_{n-1}^*)$; i.e. the scenario distribution remains the same. Both of these conclusions correspond to what should be expected from intuition.

As can be seen from the above, the crux of the argument is variance conservation: As the conditional distribution of X_n^* gets tighter, the distribution of θ_n^* must become correspondingly more flat.

It will be observed that the equations (2.6) can be represented in state space form as

$$\theta_n^* = \theta_{n-1}^* + \varepsilon_n \,,$$
$$X_n^* = \theta_n^* + Z_n \,,$$
$$n \geq 1 \,, \theta_0^* = \mu_0 \,,$$

(2.7)

where $\{\varepsilon_n\}$ and $\{Z_n\}$ are sequences of independent nonstationary normal random variables such that $\varepsilon_n \sim N(0, \sqrt{a_n - a_{n-1}}\sigma_0)$ and $Z_n \sim N(0, \sqrt{1 - a_n}\sigma_0)$. At time n the state θ_n^* and the corresponding distribution $q(X_n^*|\theta_n^*)$ will be revealed, whereas X_n^*, the true scenario, will remain unknown, unless $q(X_n^*|\theta_n^*)$ degenerates to a single point. The state equations (2.7) could be looked at as describing the possible realizations of the system which are consistent with the given initial distribution $q(X_0^*|\theta_0^*)$ and the information curve $\{a_n\}$. Also they provide us with a tool for simulating the system and for generating the transition probabilities needed in the dynamic programming algorithm.

2.2 The fitting of discrete scenario distributions

We want to insert the relationships derived in the preceding subsection into a dynamic programming framework. This requires a finite discrete set of scenarios, and in order to use the continuous conjugate distribution approach just outlined, it is necessary to discretize.

Assume that M production profile scenarios are given with initial probabilities $\pi_1, \pi_2, \ldots, \pi_M$. The distribution is assumed to be symmetric so that $\pi_j = \pi_{M-j+1}$ where $j = 1, 2, \ldots, (M+1)/2$ in case M is odd and $j = 1, 2, \ldots, M/2$ for an even M. We represent this distribution by the standard normal distribution and assign intervals $I_j = [c_j, d_j]$ such that $c_{j+1} = d_j$ and such that $\Pr(X \in I_j) = \pi_j$ for $X \sim N(0, 1)$. For example, $c_1 = -\infty$ and d_1 is determined so that $\Pr(X \leq d_1) = \pi_1$. Next, $c_2 = d_1$, and d_2 is determined so that $\Pr(c_2 \leq X \leq d_2) = \pi_2$. These intervals $I_j, j = 1, \ldots, M$ represent the possible values of the state of our variables $\{X_n\}$ and $\{\theta_n\}$. The transition probabilities,

$$q_{ij}^{(n)} = \Pr(X_n = j|\theta_n = i) = \Pr\left(X_n^* \in [c_j, d_j]|\theta_n^* \in [c_i, d_i]\right), \quad (2.8)$$

and

$$p_{ij}^{(n)} = \Pr(\theta_n = j|\theta_{n-1} = i) = \Pr\left\{\theta_n^* \in [c_j, d_j]|\theta_{n-1}^* \in [c_i, d_i]\right\}, \quad (2.9)$$

can be estimated by simulations from a large number of realizations from
(2.7) by

$$\hat{q}_{ij}^{(n)} = \frac{\#\{X_n^* \in I_j, \theta_n^* \in I_j\}}{\#\{\theta_n^* \in I_i\}}, \qquad (2.10)$$

and similarly for $p_{ij}^{(n)}$.

In the numerical computations in this paper we have chosen a cruder
approximation given by

$$
\begin{aligned}
q_{ij}^{(n)} = \Pr(X_n = j|\theta_n = i) &\approx \Pr\left(X_n^* \in [c_j, d_j]|\theta_n^* = m_i\right) \\
&= \Phi\left(\frac{d_j - m_i}{\sqrt{1 - a_n}}\right) - \Phi\left(\frac{c_j - m_i}{\sqrt{1 - a_n}}\right),
\end{aligned}
\qquad (2.11)
$$

where Φ is the cumulative distribution for the standard normal density, m_i
is the median for the interval I_i and is determined by

$$\Pr(c_i \le X \le m_i) = \Pr(m_i \le X \le d_i) = \frac{\pi_i}{2}. \qquad (2.12)$$

Similarly an approximation (not depending on simulations) for $p_{ij}^{(n)}$ is given
by

$$
\begin{aligned}
p_{ij}^{(n)} = \Pr(\theta_n = j|\theta_n = i) &\approx \Pr\left(\theta_n^* \in [c_j, d_j]|\theta_n^* = m_i\right) \\
&= \Phi\left(\frac{d_j - m_i}{\sqrt{a_n - a_{n-1}}}\right) - \Phi\left(\frac{c_j - m_i}{\sqrt{a_n - a_{n-1}}}\right).
\end{aligned}
\qquad (2.13)
$$

2.3 The dynamic programming framework

In an ordinary dynamic programming algorithm with states numbered by
$i = 1, \ldots, M$ the optimal value function $V_n, n = 0, \ldots, N$, at stage n
satisfies the recursive equation

$$V_n(i) = \max_d \left[C_n(i, d) + \beta \sum_{j=1}^{M} p_{ij}^{(n+1)}(d) V_{n+1}(j) \right], \qquad (2.14)$$

where $C_n(i, d)$ is the expected immediate return at stage n in state i and using decision d. In our context two decisions d are possible, corresponding to
a continuation of production or abandoning of the field. Further, $p_{ij}^{(n+1)}(d)$
is the $(n + 1)$-step transition probability from state i to state j and β is a

discount factor. At stage N we assume $V_N(i) = \max_d(0, C_N(i,d))$. An optimal policy can now be determined from (2.14) using standard algorithms. In our context the dynamics of the system is described both in terms of the center scenario θ_n representing the state with transition probabilities $p_{ij}^{(n)}$ given by (2.13) and in terms of the scenario distribution $q_{ij}^{(n)}$ given by (2.11). The scenario distribution enters through the immediate return $C_n(i,d)$. In fact we compute the value of the production by averaging over the production numbers for the scenarios, where each scenario is weighted in accordance with its probability of occurrence, i.e. according to $q_{ij}^{(n)}$. With a unit price of 1 and operating cost $D_n(i,d)$ at stage n, $C_n(i,d)$ is given by

$$C_n(i,d) = \sum_{j=1}^{M} Q_n(j,d)q_{ij}^{(n)} - D_n(i,d), \tag{2.15}$$

where $Q_n(j,d)$ is the production quantity for scenario j at time n using decision d.

The approximations introduced by the discretization and the formulae (2.11) and (2.13) may distort the expectations and variances so that we are no longer guaranteed that $E(X_n)$ and $\mathrm{Var}(X_n)$ are conserved. This must be taken into account by properly normalizing, since otherwise the corresponding expected production $E(Q_n|\theta_0)$ at stage n will depend on the manner in which the uncertainty is resolved.

To avoid this artefact the following scaling procedure is proposed: If we neglect the dependence on d in the original set-up, then for $\theta_n = i$ and $q_{ij}^{(n)} = \Pr(X_n = j|\theta_n = i)$, a production of

$$Q_i^{(n)} = \sum_{j=1}^{M} Q_n(j)q_{ij}^{(n)} \tag{2.16}$$

is assigned by weighting production quantities $Q_n(j)$ for the scenarios referred to in (2.15). If we let $P_i^{(n)} = \Pr(\theta_n = i|\theta_0)$, then $P_i^{(n)}$ can be computed recursively from

$$P_i^{(n)} = \sum_{k=1}^{M} P_k^{(n-1)} p_{ki}^{(n)}, \tag{2.17}$$

where $P_i^{(0)} = 1$ for $i = \theta_0$ and zero otherwise. The corresponding expected value for the production as seen from time 0 is

$$E_p(Q_n) = \sum_{i=1}^{M} P_i^{(n)} Q_i^{(n)}. \tag{2.18}$$

This may deviate from the "true" expected production at stage n given by

$$E_\pi(Q_n) = \sum_{j=1}^{M} \pi_j Q_n(j).$$ (2.19)

However, by introducing the scaling factor $E_\pi(Q_n)/E_p(Q_n)$ in (2.16) the correct expected production is obtained. This means that the immediate return $C_n(i,d)$ of (2.15) should be replaced by (d is neglected)

$$C_n(i) = \frac{E_\pi(Q_n)}{E_p(Q_n)} \sum_{j=1}^{M} Q_n(j) q_{ij}^{(n)} - D_n(i).$$ (2.20)

3. Two examples

We illustrate our method by two examples. The first example concerns the evaluation of an oil field under uncertainty, and we show that a deterministic model may lead to a substantial under-estimation of the value of the field. The second example shows that the choice of platform type may depend on how uncertainty is modelled, and that a non-optimal decision could be taken if resolution of uncertainty is not included.

3.1 Evaluation of an oil field under uncertainty

The calculations are based on artificial production data, but we have chosen them so that they closely resemble a practical case. We are interested in evaluating the value of the field as seen from time 0. Five possible production scenarios have been imagined. They are shown in Table 1 and profiles 1 to 5 are given initial probabilities 0.1, 0.2, 0.4, 0.2, and 0.1, respectively. The production profiles all have a common plateau production of 4 units a year, but the length of the plateau varies from profile to profile. Once the decline starts, the decline rate is the same for all profiles.

The price has been set to $15 per barrel, which, at an exchange rate of 6.5, gives $15 \cdot 7.33 \cdot 6.5 = 711.75$ NOK (Norwegian crowns) per Sm^3 (standard cubic meter) resulting in a yearly income of approximately 2.85 billion NOK. We use a flat oil price and fixed operating cost for each year. Discounting has been included at a yearly rate of 7%, so that $\beta = 1/1.07$ in (2.14). On the other hand taxation and inflation have been removed. This is a simple model, but it can easily be generalized in various directions. The simplicity of the model makes it easier to separate out and quantify the effect of the uncertainty resolution.

Table 1: Production profiles in mill. Sm^3/year.

Year	Profile				
	1	2	3	4	5
1	4	4	4	4	4
2	4	4	4	4	4
3	3	4	4	4	4
4	2	4	4	4	4
5	1	3	4	4	4
6		2	4	4	4
7		1	3	4	4
8			2	4	4
9			1	3	4
10				2	4
11				1	3
12					2
13					1

Four different models have been considered for the reduction in uncertainty. These are *linear reduction*, where the uncertainty is reduced by the same amount each year over the first eight years; *reduction relative to the decline point*, where 50% is reduced linearly over the first six years and then 30% and 20% over the next two years; *immediate uncertainty resolution*, where all uncertainty is revealed in the first step, and finally the *deterministic model*, where the uncertainty is not taken into account. For the last case the five production profiles and the uncertainty description is replaced by a single "expected" production profile obtained by weighting the 5 original profiles with the initial probabilities. All subsequent calculations are based on this imaginary profile.

The dynamic programming algorithm corresponding to (2.14) is given by

$$V_n(i) = \max_d \left[C_n(i,d) + \beta I(d) \sum_{j=1}^{M} p_{ij}^{(n+1)} V_{n+1}(j) \right], \qquad (3.1)$$

with $p_{ij}^{(n)}$ given by (2.13). We have included an indicator variable $I(d)$ in (3.1), where $I(0) = 0$ and $I(1) = 1$, and where $d = 1$ and $d = 0$

Table 2: Expected discounted present values in mill. NOK.

		Uncertainty resolution		
Operating cost	Linear	Relative to decline	Immediate resolution	Deterministic
0	16954	16954	16954	16953
500	13432	13407	13514	12916
1500	6850	6749	7146	6247
2000	3917	3749	4315	3464
2500	1314	1196	1730	1079

correspond to a continuation of the production and to abandoning the field, respectively. Abandonment is a nonreversible decision. In our model, once production has stopped, it cannot be started again.

For $d = 0$ the immediate return is given by $C_n(i, d) = 0$. For $d = 1$ and with a price of 711.75 NOK per Sm^3

$$C_n(i, 1) = \frac{E_\pi(Q_n)}{E_p(Q_n)} \sum_{j=1}^{M} Q_n(j) q_{ij}^{(n)} \cdot 711.75 - D, \qquad (3.2)$$

where D is the fixed operating cost, and where $q_{ij}^{(n)}$ is computed using (2.11).

The problem at hand is an optimal stopping problem. In the hypothetical case of $D = 0$, it will be optimal to continue producing until the resources are exhausted, i.e. until the end of the production profile is reached. The expected value of the field is obtained by multiplying the expected production profile by the unit price, and we should get the same result irrespective of the way the uncertainty is revealed, since no decision is involved. This is confirmed from the first line of Table 2.

The other extreme situation is where D is very large compared to the yearly production income of 2.85 billion per year at the plateau. The last line of the table treats such a case with $D = 2.5$. In the deterministic case the production is stopped as soon as the expected production level reaches 2.5, whereas if the uncertainty is resolved immediately (or before the time of decline for the shortest profile), the production is continued as long as the production for the profile we are on, exceeds 2.5. For the deterministic approach the production stops too early in case we are on a long profile and too late in case we are on a short profile, and it is seen from the last line of Table 2 that the expected values are almost twice as large in the case of immediate resolution of uncertainty.

The same pattern, with values being highest where the information is revealed most quickly, is observed for the intermediate cases $D = 2$, 1.5 and 0.5, although with progressively smaller differences. Thus the output of the algorithm confirms the qualitative beliefs of our intuition. In addition these beliefs are quantified numerically.

3.2 Application to the choice of platform design

The calculations are made for the field Draugen, which is a relatively small oil field off the coast of Norway. The owners have considered a number of development plans. Two main alternatives are: a standard solution with an immovable platform or a movable platform (which may be rented).

The production and cost data are given in Table 3. Wood-MacKenzie (1988) has been used as a source. In the last column the capital cost associated with the immovable platform has been entered.

Wood-MacKenzie has used a deterministic method based on net present values, where the field is abandoned when the income due to the production data of Table 3 equals the expenses. For the immovable platform alternative they find an after-tax expected present value of 1664 million NOK discounted to the end of 1987 using a discount rate of 12% and an exchange rate of NOK $6.20 = \$1$ from 1988 onwards. The oil price has been taken to be \$16 per barrel in 1988, \$18 in 1989, \$20 in 1990 with a 6% per annum increase thereafter. Moreover, a future inflation of 5% per annum is assumed from 1989. All figures are in current \$ or NOK. If the production profile of Table 3 is interpreted as an expected profile obtained by averaging over several possible production scenarios, their method will give a too low value, as is explained in the introduction.

To be able to focus more sharply on the main aspects of the problem we will make some simplifications. First, we neglect possible effects of taxation, and hence all of our calculations are before tax. In addition we assume that the gas has no economic use. Indeed, the value of the small gas reserves of Draugen is very much dependent on a future transportation system. Further, it is assumed that the immovable platform remains stationary after production has stopped, so that (contrary to Wood-MacKenzie) no removal cost is added at the end of the production period. The interest rate is set to 12% to obtain the 7% real interest rate used by Norwegian authorities. The exchange rate, operating and capital cost are the same as those used by Wood-MacKenzie. However, we assume that the oil price

Table 3: Wood-MacKenzie production data for Draugen.

Year	Oil production 10^3 barrels per day	Gas Billion cubic feet	Operating cost mill. NOK	Capital cost mill. NOK
1988				
1989				407
1990				2316
1991				3287
1992	25		495	2254
1993	90		1026	845
1994	90		1077	563
1995	90		1131	344
1996	90		1188	
1997	90		1247	
1998	80		1310	
1999	75		1375	
2000	65		1444	
2001	55		1516	
2002	45		1592	
2003	40		1671	
2004	35	10	1880	
2005	30	30	2231	
2006	25	30	2342	
2007	20	30	2459	
2008	18	30	2582	
2009	16	30	2712	
2010	13	5	2434	

stays fixed at \$15 per barrel in 1988 money. Under these assumptions a deterministic model yields a value of 2821 million NOK.

It remains to introduce uncertainty into the model. It is natural to assume that the production profile of Table 3 is the most likely one, and we assign an initial probability of 0.4 to a slightly shortened version of it. Four other production profiles are introduced. This is done in such a way that the profile of Table 3 is close to being the expected production profile of the stochastic model. The production profiles with initial probabilities are given in Table 4.

In order to simplify, the uncertainty is supposed to be resolved at a uniform linear rate starting with production and being completely resolved during a 5-year period irrespective of the production profile. In the absence of a specific model this a reasonable way to proceed. The time of abandonment and the value of the field using the immovable alternative

Table 4: Production profiles Draugen, 10^3 barrels per day.

Year	Production profile				
	1 $\pi_1 = 0.1$	2 $\pi_2 = 0.2$	3 $\pi_3 = 0.4$	4 $\pi_4 = 0.2$	5 $\pi_5 = 0.1$
1992	25	25	25	25	25
1993	60	80	90	90	90
1994	50	70	90	90	90
1995	40	60	90	90	90
1996	30	50	90	90	90
1997	20	40	90	90	90
1998	10	30	80	90	90
1999		20	70	90	90
2000		10	60	90	90
2001			50	80	90
2002			40	70	90
2003			30	60	90
2004			20	50	90
2005			10	40	80
2006				30	70
2007				20	60
2008				10	50
2009					40
2010					30
2011					20
2012					10

Table 5: Expected present values in mill. NOK. Draugen.

	Stochastic	Deterministic
Immovable	3593	2821
Movable	3894	2824

is evaluated using the dynamic programming algorithm and the data of Tables 3 and 4. As compared to the deterministic model, the value rises with more than 700 million NOK to 3593 million NOK. An even greater rise would have been observed if the uncertainty had been resolved more quickly as in Figure 2 for example.

For the movable alternative it is assumed that is is only possible to rent the platform for an annual rent. Then the capital cost of Table 3 is replaced by a yearly renting cost paid from the year 1992 onwards and escalated with the rate of inflation until production stops.

A main problem is to find the rent which makes the two platform alternatives equivalent. If we use the expected production profile and a deterministic model, it is not difficult to perform this calculation. It is found that a yearly rent of 2.22 billion NOK paid from the year 1992 onwards and escalated with the rate of inflation gives approximately the same expected present values for the two alternatives (cf. Table 5).

With the annual cost of 2.22 billion NOK we now introduce the stochastic model defined in Tables 3 and 4. We observe from Table 5 that the values of both alternatives are increased with more than 20%, and that the movable platform alternative is clearly better, the difference being 0.3 billion NOK. Actually the difference in our particular case is deflated because the important shut down decision is on the average more than 10 years from the start of the investment, and therefore the difference between the two models is discounted to a considerable degree. In addition the uncertainty in production has a negative effect on the expected present value in this model. This follows since the value of the field is an increasing concave function of the plateau length.

Although the difference is not exceedingly large, it illustrates our point well. If uncertainty is neglected, one could easily end up with choosing the wrong investment.

4. A simple tax system

In many countries natural resource exploitation is taxed differently from other industries. In Norway the oil producing industry faces an average tax rate of about 80%. An interesting question is whether the pre-tax best development plan will be changed by introducing tax in the model.

It is difficult to introduce taxation into dynamic programming models. Typically tax rules require more state variables. Any tax rule independent of past events is simple to implement. On the other hand tax rules dependent on history such as depreciation, accumulated losses etc. necessitate more state variables.

Another special feature of petroleum taxation is the ringfence effect. The ringfence around a firm's petroleum activities means that losses can not be deducted from income stemming from non-petroleum related business. This in turn implies that losses must be carried forward in many situations. Low oil price might give big oil companies losses. To isolate this effect on our decision problem we have constructed the following very simple tax system:

(1) There is a ringfence around any field, hence any investment will imply that losses must be carried forwards.
(2) Investments, operating costs and accumulated losses are deductible from the annual profit (immediate depreciation for tax purposes).
(3) The tax rate is 80%.
(4) Only the nominal value of losses is carried forwards.

This simple tax model can be implemented into our framework by introducing a new state variable k describing accumulated loss. The variable k is assumed to be integer valued obtained by discretizing the loss in appropriate units.

We need some notation. Referring to equation (2.20) we denote by

$$R_n(i) = \frac{E_\pi(Q_n)}{E_p(Q_n)} \sum_{j=1}^M Q_n(j) q_{ij}^{(n)} \tag{4.1}$$

the scaled anticipated production in resource state i. The price is assumed to be constant and is given by s. As before D_n is the investment and operating cost. It is assumed to be independent of i. Finally, we denote by L_k the accumulated loss in tax state k and by t the tax rate.

The after tax profit $\Pi_n(i,k)$ in period n can now be written as

$$\Pi_n(i,k) = R_n(i) \cdot s - D_n - \max(R_n(i) \cdot s - D_n - L_k, 0)t \tag{4.2}$$

The tax state k' in the next stage is given by (we have adjusted for an inflation rate of size r)

$$k' = \text{integer} \left[\frac{\max(L_k - R_n(i) \cdot s + D_n, 0)}{(1+r)} \right] \tag{4.3}$$

If we let $V_n(i,k)$ be the optimal value function for resource state i and tax state k, the following dynamic programming recursion is obtained.

$$V_n(i,k) = \max \left[0, \Pi_n(i,k) + \beta \sum_{j=1}^M p_{ij}^{(n+1)} V_{n+1}(j,k') \right]. \tag{4.4}$$

This formula should be compared to the pre-tax formula (2.14).

Using this model gives the following conclusion: It pays to substitute operating cost for investment cost in situations where the losses are reduced when carried forwards. Further, an upfront investment can be lost for particularly poor production outcomes, since then this loss can not be deducted from future income. All of this is well known and implies that flexible operating costs are preferred (see e.g. Lund (1987) and the references therein).

In a model where all costs are operating costs this tax system has little impact on the optimal policy. For the examples considered by us the only relevant situation is when the annual profit for a certain state and period is negative, but where the probability of increased production in the next period prevents a shut down. In this situation the after-tax optimal decision will tend to give an earlier shut down, in particular in cases where one knows quite early that there is a high probability for a short production profile.

We have investigated the example of Section 3.1 from this point of view. The production data are given by Table 1, and we use both a linear resolution of uncertainty and a resolution of uncertainty relative to decline. We employ a discount rate of 7%. The relevant pre tax results are given by Table 2.

The effect of taxation will be largest when the operating cost is very high. We have therefore chosen an operating cost of 2.82 billion NOK, which is larger than all of the operating costs used in Table 2, and which is very close to the yearly production income of 2.85 at the plateau. The pre- and after-tax expected present values are given in Table 6. For the after-tax computation an inflation rate r of 5% has been used. It is seen that the taxation leads to a small loss in total expected value. This is due to an earlier shut down as explained above.

Table 6: Expected present values in mill. NOK before and after tax with an operation cost of 2820 mill. NOK.

Uncertainty resolution	Before tax	After tax	Tax payment	Loss in total expected value
Linear	109.22	21.48	86.10	1.64
Relative to decline	64.80	12.59	50.44	1.67

5. Uncertain prices

In this paper our main concern has been the production model with uncertain production scenarios and with an incorporation of gradual reduction of uncertainty. Uncertainty in price has been neglected. Here we introduce a simple price model; the main point being to show how uncertainty can be included and how its effects can be measured. More complicated price models can be treated analogously.

We assume that the price $\{S_n\}$ is discretized and that is moves among K possible states according to a Markov chain. This does not necessarily imply that future prices only depend on the present price, since models with higher order time lags can be implemented as Markov chains by vectorization and extension of the state space.

The state variable now consists of two parts, production and price, and corresponding to (2.14) we get the dynamic programming recursion

$$V_n(i, k) = \max \left\{ C_n(i, k, d) + \beta \sum_{j=1}^{M} \sum_{m=1}^{K} p_{ij,km}^{(n+1)}(d) V_{n+1}(j, m) \right\}, \quad (5.1)$$

where $V_n(i, k)$ is the optimal value function for production state $\theta_n = i$ and price state $S_n = k$ at time n, and where

$$p_{ij,km}^{(n+1)} = \Pr(\theta_{n+1} = j, S_{n+1} = m | \theta_n = i, S_n = k). \quad (5.2)$$

The immediate return $C_n(i, k, d)$ is given by

$$C(i, k, d) = \frac{E_\pi(Q_n)}{E_p(Q_n)} \sum_{j=1}^{M} Q_n(j, d) q_{ij}^{(n)} s_k - D_n(i), \quad (5.3)$$

with s_k being the actual value of the price in state k.

The algorithm simplifies if we assume independence between production and price (which for oil production can be done to quite a good approximation). Then $p_{ij,km}^{(n+1)} = p_{ij}^{(n+1)} r_{km}^{(n+1)}$ where $p_{ij}^{(n+1)}$ has the same interpretation as in (2.14), and where $r_{km}^{(n+1)} = \Pr(S_{n+1} = m | S_n = k)$. For purposes of illustration we have chosen to model $\{S_n\}$ as obtained by discretizing a geometric Brownian motion. This means (Stensland and Tjøstheim 1989) that S_n satisfies the equation

$$\ln(S_n) - \ln(S_{n-1}) = \delta + e_n \quad (5.4)$$

Table 7: Expected discounted present values in mill. NOK. Stochastic price.

Operating cost	Uncertainty resolution		
	Linear	Relative to decline	Immediate resolution
0	16711	16711	16711
500	13233	13210	13324
1500	7134	7058	7368
2000	4631	4550	4887
2500	2580	2509	2819

where $\{e_n\}$ is a Gaussian zero-mean white noise process, and where δ and $\sigma^2 = \mathrm{Var}(e_n)$ are parameters which can be estimated from historic data on oil prices.

The continuous analogue of this model has been used in natural resource modelling (Brennan and Schwartz 1985). More complicated discrete time models are presented in Stensland and Tjøstheim (1989) for the case of copper data.

Using independence and the simple Markov model based on (5.4) we obtain the results of Table 7 for the example in Section 3.1. The first line of the table has not been completely calibrated to the first line of Table 2, but the message emerging from these two tables is clear. We observe that the value of the investment opportunity increases when uncertainty in price is introduced. This is consistent with standard arguments in option pricing theory: The value of the shut down option increases with increased uncertainty (see e.g. Brennan and Schwartz 1985). We also observe that the differences between the three uncertainty resolution methods are less when price uncertainty is introduced, although the relative ranking between them is retained.

6. Some extensions and generalizations

Our work can be generalized in a number of directions, and we will mention a few of them briefly. We concentrate on the model in Sections 2 and 3.

a) The generation of scenarios and physical models

The assignment of probabilities to scenarios is a difficult task and must be undertaken by experts. In case there is a physical model for generating the scenarios from basic geological and geophysical parameters one may instead assign probabilities to the parameter values. In fact several models exist for linking production to the reservoir distribution and to the parameters determining it. An example is given in Asheim (1986).

b) Noise in the production profiles

The production profiles we have used with a clearly defined plateau and decline phase are idealized. Real production profiles have a much more rough appearance. We refer to Stensland and Tjøstheim (1988). Such irregular features can be introduced by including a separate stochastic process describing production profile noise due to production stops etc. Such a noise process may cause a lengthening or shortening of the time period to decline and abandonment, and the model must be adjusted accordingly.

c) Other sequential decision problems

In section 3 the only decision problem involved is that of abandonment. Clearly there are a number of other decision problems which are of interest and which, in principle, may be included in the model. Some of these may in fact be considerably more sensitive to model formulation than the abandonment problem. Some examples are: Determining the number of production wells and the size of the platform, use of injection and use of twostep investments when satellite fields are present.

d) Improvement of the mathematical model of Section 2

Several production levels could be introduced at the plateau by means of a probability mixture model with a set of a priori probabilities generating the possible levels and corresponding scenario distributions for each level. The scenario distribution could be skewed, e.g. a gamma distribution.

Further, alternative weighting procedures could be considered. Separate uncertainty resolutions could be prescribed for various parameters such as production level, time of decline or geophysical/geological parameters determining these. Finally, it may be useful to introduce correlation if several variables are present.

References

Aaheim A. (1986): "Depletion of large gas fields with thin oil layers and uncertain stocks", Preliminary report, Central Bureau of Statistics, Oslo, Norway.

Arps J.J. (1956): "Estimation of primary oil reserves Petroleum Transactions", *AIME* 207: 182–191.

Asheim H. (1986): "Effect of production interventions estimated using exponential decline curves", Report, Norwegian Technical University, NTH, Trondheim, Norway.

Bjørstad H., T. Hefting, and G. Stensland (1989): "A model for exploration decisions", *Energy Economics* 11: 187–196.

Brennan M.J., and E.S. Schwartz (1985): "Evaluating natural resource investments", *Journal of Business* 58: 135–157.

Burdett K., and T. Vishwanath (1988): "Declining reservation wages and learning", *Review of Economic Studies* 55: 655–666.

De Groot, M.H. (1970): *Optimal Statistical Decisions*, New York: McGraw Hill.

Deshmukh S.D., and S.R. Pliska (1980): "Optimal consumption and exploration of nonrenewable resources under uncertainty", *Econometrica* 48: 177–200.

Gilbert R.J. (1979): "Search strategies and private incentives for resource exploration", *Advances in the Economics of Energy and Resources.* JAI Press 2: 149–169.

Hoel M. (1978): "Resource extraction, uncertainty and learning", *Bell Journal of Economics* 9: 642–645.

Kemp M. (1976): "How to eat a cake of unknown size", in *Three Topics in the Theory of International Trade*, Amsterdam: North Holland.

Loury G.S. (1978): "The optimal exploitation of an unknown reserve", *Review of Economic Studies* 45: 621–636.

Lund D. (1987): "Investment, taxes, and uncertainty with applications to the Norwegian petroleum sector", Memorandum, Department of Economics, University of Oslo.

Norwegian Oil Directorate (1987): *Future Outlook* (in Norwegian), Stavanger, Norway.

Pindyck R.S. (1980): "Uncertainty and exhaustible resource markets", *Journal of Political Economy* 88: 1203–1225.

Raiffa H., and L.R. Schlaifer (1961): *Applied Statistical Decision Theory*, Harvard University, Boston.

Roberts K., and M.L. Weitzman (1981): "Funding criteria for research, development and exploration projects", *Econometrica* 49: 1261–1288.

Rothschild M. (1974): "Searching for the lowest price when the distribution of price is unknown", *Journal of Political Economy* 82: 689–712.

Olsen T., and G. Stensland (1988): "Optimal shut down decisions in resource extraction", *Economic Letters* 26: 215–218.

Olsen T., and G. Stensland (1989): "Optimal sequencing of resource pools under uncertainty", *Journal of Environmental Economics and Management* 17: 83–97.

Stensland G., and D. Tjøstheim (1988): "Production uncertainty (in Norwegian)", CMI report No. 30152–1, Centre for Petroleum Economics, N-5036 Fantoft, Norway.

Stensland G., and D. Tjøstheim (1989): "Optimal investments using empirical dynamic programming with application to natural resources", *Journal of Business* 62: 99–120.

Wood-MacKenzie & Co. (1988): *North West Europe Service*.

Author index

Subject index